ECONOMIC AND
SOCIAL DEVELOPMENT

A Process of Social Learning

Edgar S. Dunn, Jr.

Published for
RESOURCES FOR THE FUTURE, INC.

by
THE JOHNS HOPKINS PRESS
Baltimore and London

Resources for the Future is a nonprofit corporation for research and education in the development, conservation, and use of natural resources and the improvement of the quality of the environment. It was established in 1952 with the cooperation of the Ford Foundation. Part of the work of Resources for the Future is carried out by its resident staff; part is supported by grants to universities and other nonprofit organizations. Unless otherwise stated, interpretations and conclusions in RFF publications are those of the authors; the organization takes responsibility for the selection of significant subjects for study, the competence of the researchers, and their freedom of inquiry.

This book is one of RFF's regional and urban studies, which are directed by Lowdon Wingo. Edgar S. Dunn, Jr., is a senior research associate with Resources for the Future. The index was compiled by Helen Eisenhart.

RFF editors: Henry Jarrett, Vera W. Dodds, Nora E. Roots, Tadd Fisher.

Preface

THIS BOOK IS A HAPPENING. It grew, unpremeditated, out of the anomalous experiences that accompany social science research. The step-wise process of individual learning brought me here, aided and abetted by Albert Hirshman's "principle of the hiding hand." I was protected by my ignorance and did not know in advance the difficulties and troubles that lay in store. I have been rewarded by an adventure in learning, of which I hope this volume represents only the first stage.

This is a book about human problem solving seen as a process of social learning. It grew out of a process of problem solving and a record of its evolution would, itself, form a good illustration of a number of aspects of the process here described. Several colleagues have taken note of this fact. It is hoped, however, that the organization of this volume bears only minor traces of this process, for such a representation of the learning path would not serve best the requirements of exposition and communication.

The theme is the following: Evolution, biological and social, is a learning process. Phylogenesis is a learning process that has bred animal learning and human learning leading to social learning. Social learning incorporates many characteristics of its antecedent learning processes but exhibits many unique characteristics of its own. Some of these are becoming evident. Others will require the advance of social science to uncover. Many of the most critical human social problems of our day are developmental problems that can only be successfully approached through the medium of some understanding of this learning process. However, the conventional metaphors and methods of economics and the other social sciences are deterministic models that cannot be stretched to accommodate adequately the reality of social development. They are adequate to deal with only a subset of the problems associated with social change. Even then, their application should be viewed as taking place under the control of a more comprehensive view of the process.

The credits that are appropriate for this preface are almost without limit. At the head I must acknowledge my colleagues at Resources for the Future. Joseph Fisher, Michael Brewer, Harvey Perloff, and Lowdon Wingo have been most generous with their understanding and unstinting in their support of so unconventional an undertaking. In addition, Irving Hoch, José-Ramón Lasuén, Gordon Cameron, Jay Polach, and Edwin Haefele have not only read the manuscript but have spent countless hours discussing issues with me during the successive rounds of drafting.

Among others who reviewed earlier drafts I am especially indebted to Otis Dudley Duncan, George Stolnitz, David Rosenblatt, Samuel Klausner, Ruth Mack, and Harold Sackman. I am also grateful to Bernhard Rensch, Theodosius Dobshansky, and E. Lloyd DuBrul, who reviewed the first draft of the section on biological evolution.

Beyond these, who must be accorded special recognition, I am indebted to that host of honest minds who brought antecedent scholarship to the point where such an image as is offered here could emerge. Many of them have been cited in one way or another in the text, but by no means all. I will soon be indebted to the continuing procession of scholars who through their own work will extend what is valid and reveal what is weak.

There must be ideas contained in these pages that should be attributed to others but are not. I reached the point in talking with fellow scholars and in reviewing a voluminous literature, where I was not always certain where my ideas began and others left off. Reasonable care with attribution may not have been sufficient. I have undoubtedly also expressed ideas that already exist in a vast literature undiscovered by me. There is a natural duplication of ideas when people begin working in common channels. But my concern is not for personal credits for specific ideas or establishing the elements of novelty in my presentation as compared with the thoughts of others. My only interest is in sharing with the reader a set of ideas and an experience in learning that has given new meaning and force to my career as a professional social scientist. I am anxious to learn whether that experience is valid for others as well. I fear chiefly that my perception and my expression may not be extravagant enough, may not wander far enough beyond the limits of my experience and scholarship, to be adequate to the social science paradigm I am beginning to perceive.

In addition to the many from whom I have received professional support, two have given personal support of a kind that fulfilled a necessary condition for this effort to be brought to completion—my wife, Lillian, and Thaxton Springfield. Unfortunately, it is not possible to convey to anyone else the significance of their contribution. Their credits have to do with lives, not books. This is a prior and higher order of creativity.

JUNE 1970   EDGAR S. DUNN, JR.

To say that man accomplishes nothing but that to which his endeavors are directed would be a cruel condemnation of the great bulk of mankind and their families. But, without directly striving for it they perform all that civilization requires, and bring forth another generation to advance history another step. Their fruit is, therefore, collective; it is the achievement of the whole people. What is it, then, that the whole people is about, what is this civilization that is the outcome of history, but is never completed: We cannot expect to attain a complete conception of it; but we can see that it is a gradual *process,* that *it involves a realization of ideas* in man's consciousness and in his works, and that it takes place by virtue of man's *capacity for learning,* and by experience continually pouring upon him *ideas* that he has not yet acquired.

... we are all putting our shoulders to the wheel for an end that none of us can catch more than a glimpse at—that which the generations are working out. But we can see that the *development of embodied ideas* is what it will consist in.—

CHARLES S. PEIRCE, 1893 (italics added).

Contents

Part I

INTRODUCTION

I   A Growing Concern with More General Approaches to Prediction and Planning

THE IDEAS THAT EMERGE in a systematic progression between these covers have both a special and a general orientation. Their special orientation derives from the fact that the author is an economist concerned with the problems of regional and national economic development. The student who addresses himself to these problems is continuously confronted with restrictions imposed by traditional concepts and methods.

The pages that follow represent an economist's attempt to reach into a less constrained domain of concept in the interest of gaining new insights into the processes of social change and the methods by which they may be studied. In so doing, he has been led to deal with a set of ideas of much more general scope than those usually associated with economics or with regional or national economic development. Through a portal of special interest he enters into a domain of general principle that appears to be equally relevant to the exercise of other social science disciplines and to other special aspects of the process of social change.

This is much more than a book on regional and national development, therefore, and it may claim the interest of all social scientists regardless of their special concerns. Different disciplines are working their way into the same conceptual domain through diversely placed portals. Nevertheless, the perspective given to this search by the author's "port of entry" is retained. This is dictated in part by the fact that the process of cognition requires any "idea-scape" to be viewed initially from a fixed point of reference, even though one's understanding of the elements of the scene may be enriched by different perspectives of the same topology from different points of reference. It is dictated, further, by the limitations of the author's achievements as a scholar. We proceed through the portal of economic development.

It cannot have escaped widespread notice that in recent years economists have been devoting a great deal of attention to the subject of economic "development" or "growth." Indeed, the set of problems collected under this general rubric has become a dominant concern for the current generation of economists, just as did the issue of economic stability for the previous generation.

This interest is generating considerable research and a burgeoning literature. The literature appears to have divided into two main shoots which, up to the present stage, have had little direct contact. On the one hand there is a group absorbed with issues of "regional development." Its attention has been focused on regional subsystems primarily within the advanced Western economies and its concern has been addressed primarily to improvements in the efficiency of the internal organization of a modern economy or some component thereof. On the other hand, there is a group concerned with the issues of national economic development. Their attention has been directed to the problems of the economies that have remained backward and how they can develop behavioral capacities corresponding to those of the advanced countries.

Within each group certain minor themes tend to stray away from the characteristic one and echo with some variation the principal concern of the other group. Thus, the regional analysts develop special interest in relatively "backward" regions in the advanced economies and their evaluations and prescriptions sound remarkably like those concerned with national economic development. Within the latter group there has grown an awareness that "general" development goals require "specific" implementation plans. As the national development group becomes concerned with these problems of economic organization they begin to sound strangely like the regionalists. But in the main it remains true that both groups, while displaying a similar interest in economic growth and development, have been concentrating on the process of economic change in different empirical settings and have been following distinctly different paths.

It is interesting, however, to note the growing signs of convergence as the idea content of these efforts evolves. There are indications that each of the separate intellectual shoots is moving from a more limited to a more general concern with prediction and planning. In each there also appears to be a growing concern that the conventional methods of the economist are not fully adequate for an understanding of the development problem. Our topic will unfold in more orderly fashion if each of these analytical traditions is reviewed briefly.

Evidence of the Growing Concern

Work in regional economic analysis has been dominated by a concern with homogeneous or behaviorally simple production systems. The bulk of the work has dealt with activity analysis that is regionally specific. This is especially true of the substantial volume of study and projection used to support physical planning or investment planning. Take the field of water resource planning as an example. The analysis is directed primarily to a single homogeneous system. It is a planning analysis that happens to be spatially constrained by the attributes of hydrology on the supply side and the attributes of market on the demand side and the technology that links them. Hence, the analysis is often considered to be regional analysis, but the analysis of the region is incidental to the analysis of the activity. The same can be said for a large array of public and private physical planning exercises, such as transportation planning and plant location analysis.

Recent work has led to analysis with a more primary regional orientation. This has developed in the field of community or urban planning where sets of related activities that constitute a regional bundle need to be evaluated in terms of their efficient organization and the requirements for community services. Next to physical activity planning, community and urban planning have given rise to the largest volume of expenditure and effort. The objectives of prediction and planning have been more general (and the problems of method and concept greater) by virtue of the multiple-activity, systemic nature of the region concept. The domain is explicitly regional in a more fundamental sense than is characteristic of the analysis of an activity. The attention, however, is conventionally addressed to a single region or urban system.

More recently there has emerged a growing concern with the need to know how complex regional activity sets are changing over time because (1) regional maladjustments are presumed to be creating social pathologies requiring public attention or because (2) efficient long-run planning of public and private functions requires it. In short, there is more interest in understanding the growth process of complex dynamic systems and in development planning for social systems. Accompanying this is a tendency to examine a more complex array of systems—to investigate multiactivity regions and to see them in the context of their relationships with other regions and the nation. This is the area where the more vital interests of long-run public policy seem to be engaged. Hence, the prediction and planning objectives become more closely tied to the goal of understanding processes of social change in a fundamental sense and less narrowly oriented to a single activity. True, so far the more general objectives have been served in only a limited and unsuccessful way; nevertheless, regional economic analysis is gradually moving toward a more general view of prediction and planning. Its domain of interest is also moving closer to the domain of national economic development because of the importance of seeing regional change as related to the transformation of the total systems of which the regions are a part.

The literature dealing with national economic development reveals similar tendencies. The largest part of it has dealt with the growth and development problem on the basis of very simple aggregated system concepts. Growth has most commonly been associated with the phenomenon of aggregate capital accumulation. At times, particularly when approached in the context of modern Western economies, this is cast in the framework of the dynamic Keynesian model. The Harrod-Domar extension of the Keynesian model deals with such concepts as "moving equilibria" or the gradual upward shifts of "normal levels." In the literature dealing with the development of backward economies the same general ideas occur in a somewhat different setting. There is a preoccupation with how a lagging economy can be displaced to forms of organization and activity that conform more closely to the standards set by advanced Western economies. The problem is conceived as a process of modifying the balance of capital and human resources to conform more to that exhibited by advanced economies. Capital accumulation is commonly identified as the key variable in determining the rate of development.

Net investment has occupied stage center. In the advanced country setting this has led to a concern with the techniques of keeping the savings-investment relationship in some sort of dynamic balance. In the lagging countries it has led more to a concern with the role of such elements as agriculture, taxation, inflation, investment, and international trade, considered from the viewpoint of their significance as sources of savings.

Increasingly, however, these aggregate-level simple system concepts have been extended in two principal directions. Greater attention has come to be given to the allocation of investment between developmental options. In one strand of the literature this has taken the form of a series of semidisaggregated investment guides or criteria. "Minimum capital intensity" (a concept tied to the capital-output ratio), "maximum social marginal productivity of capital," the "marginal reinvestment quotient," and "maximum employment absorption" are all put forth in a voluminous literature. (See Meier, 1966.) In another strand of the literature a more explicit recognition is taken of economic subsystems including the economic region and urban region. National development policy is often articulated in terms of the regional or other subsystem activities required for its implementation.

Concepts are also increasingly extended in another direction. There are occasional attempts to identify necessary growth sequences or "stages" in the interest of introducing a more dynamic framework and to aid the process of prediction and planning.

Several trends are emerging in professional literature. First, there is clearly a tendency to consider economic development in the context of more complex functional systems. Second, through the more intensive use of historical materials and longer-range prediction techniques, there is a tendency to try to extend the time period of relevant prediction and planning. Third, the literature on regional economic development and national economic development is moving rapidly toward some convergence: regional analysis, from a concern with simple sub-

7

systems to a consideration of their relationship with linked systems of a higher order, national economic development, from a concern with aggregated simple systems to a consideration of their relationship with linked subsystems. Both are seeking to extend their time horizons.

On the one hand, one identifies a growing need to view the activities of behavioral units as a network of linked functions making up a complex interdependent system—the latter seen as susceptible to differentiation into functional subsets. On the other hand, one identifies a growing need to understand more fully the processes by which such complex systems are transformed.

This, of course, is a grossly oversimplified characterization of the history of thought in these economic fields. Its purpose is solely to illustrate the way in which today's thinking is tending to generalize prediction and planning in order to treat more complex functional systems and more extended time perspectives. Clearly, a restive search is under way for concepts that are more relevant to interpretation of the harsh realities disclosed by experience. To date the net effect is uncertain and no clearly unified direction of effort has emerged.

But there is a much more important and revealing way in which the approach to prediction and planning is becoming more generalized. The growing concern with more general, more realistic approaches to prediction and planning is reaching out to embrace more general concept models of the structure and change of economic systems. This is clearly implied in the movement to encompass more complex functional systems and more extended time perspectives. It is through an explicit consideration of these concept models that we can get our clearest picture of the generalizing movement and the degree to which the various models meet the requirements of prediction and planning. Before turning to this task, however, we should clear up some ambiguity commonly found in the use of the terms "growth" and "development."

The Concepts of Growth and Development

The terms "growth" and "development" are often used interchangeably or imprecisely in current literature. Since a clear understanding of their meaning is necessary to the discussion in the next section, the following distinction is suggested. *Growth* implies that an activity system is increasing the scale of its social structures and the quantitative level of its activities. Population, employment, income, etc. are dimensions commonly used to reflect these changes in level or scale. *Development* implies that an activity system is transformed in the mode of its behavior. Growth is a scale concept and development a behavioral concept. We consider growth and development to be the positive dimensions of changes in scale and behavioral mode. Growth implies an increase in scale, development an increase in the complexity of behavior. Implicit in these concepts are their opposites. The concept of growth implies the concept of decline. The concept of development implies the concept of behavioral regression. Social change embraces the positive and negative dimensions of both changes in scale and changes in behavioral mode.

At the outset, these terms need take on no normative connotations. The association of the concept of development with increases in behavioral complexity requires no more justification for the moment beyond the fact that the emergent process we know as biological and social evolution has been characterized by increases in system complexity. Over the long run these, in turn, have been associated with improvements in adaptive capacity. Normative implications for the use of these terms will be developed at later points in the discussion.

Growth and development are not necessarily perfectly correlated in a finite time span. In the biological realm an evolutionary increase in the behavioral complexity of a species of organisms almost always implies an increase in the size of the organism. However, numerous increases in size have been identified that do not imply increases in complexity during the same evolutionary span. Similarly, increases in the scale or activity levels of a social system may be achieved through the aggregation of existing behavioral modes made possible by the availability of unexploited free resources. Thus, growth is not a sufficient test of development.

In the social realm the direct correlation is less certain because the organization of a subsystem can be modified on a part-by-part basis. For example, a social activity system like an urban region that possesses any internal subsystem complexity may simultaneously modify the behavioral mode of some components while experiencing, for the same or different reasons, a decline in the aggregate size of other components. The net effect for a given system might, therefore, be a decrease

9

in the size of some total system dimensions accompanied by regional development.

In the end, of course, growth or changes in scale must rest upon true development (upon changes in modes of behavior either internal or external to the system) and will have to be evaluated in those terms if we can develop the skill. The most important thing about social change or urban system transformation is not the changes in scale but the changes in behavior because, ultimately, all positive changes in scale are dependent upon the latter. Furthermore, it is behavioral change in a social system that generates the special problems of adaptation that confront decision makers. The average correlation of growth and development is high over long periods but may be missing for some activity systems at any point in time.[1]

[1] It is quite common for some people to think primarily in terms of growth when they talk of regional, industrial, or national development. This is partly because they may not perceive the imperfect or lagged nature of the link between growth and development or may fail to perceive the distinction between scale and behavioral mode. It may also stem from the fact that they perceive a more direct relationship between activity levels and the value of ownership vested in resources and, therefore, the benefits that can accrue to the individual with resource claims. This point of view may betray a static-state view of reality. This habit of mind is reinforced because the safe way to increase size is through aggregation while true development runs the risk that old resource claims may be unfavorably revalued relative to the new.

If one is concerned with the growth of an activity system rather than its development and hopes to stimulate it through the aggregation of that mode of behavior, that growth can only be achieved through the beneficence of some externally linked activity. This is why the analysis of economic growth is so commonly addressed to the evaluation of the prospects inherent in these links. The conventional economic framework takes naturally to this task because of its propensity to focus upon external demand-supply or input-output relationships and to see internal change as fixed program adaptation. There are scores of research studies, ostensibly of regional development (as well as national development), that devote great energy and ingenuity to evaluating the consequences of external sources of change for the region's traditional modes of behavior. Further, those that make development a human goal or social objective characteristically find this goal consistent with growth as an objective. On the other hand, those that promote growth as a system objective may have vested reasons for seeking changes in scale without behavioral change. They may adopt a point of view and foster policies that in fact inhibit development. They will, in fact, also inhibit growth in the long run, but (a) they are not sensitive to this fact and/or (b) they have a short-term value horizon.

Common Concept Models of Structure and Change in Social Systems

We have concluded that an explicit consideration of prevailing concept models of structure and change will provide the clearest picture of the move to generalize the practice of prediction and planning. It will also set the stage for the theme we wish to develop in these pages. We turn, therefore, to a consideration of these model concepts. They fall naturally into growth models and development models. They may be combined to form more complex models.

GROWTH MODELS

In considering the phenomenon of growth, it is quite natural that attention was focused at an early point upon increases in the aggregate scale of some social system—be it a region, nation, or a differently bounded social system. The simplest concept is that which considers the system growth to represent a historical trend or a form of natural increase that can be extrapolated to form a prediction of system size at a point in the future.

If one stops at this point he has encountered nothing more than descriptive history combined with faith in the inexorability of its trend. If one is to have any confidence in such a projection of historical trend, one requires an explanation that justifies the faith. Historically, three principal modes of explanation have been offered.

Simple Growth Models. The first is the biological law of natural increase in population size. It rests on the notion that a reproducing species will increase its population naturally and that the increase will follow the form of a geometric progression with the rate of growth dependent upon the mechanisms of reproduction. Human populations are not held to be different from any biological species in this regard. But this mode of explanation standing alone does not support confidence in the extrapolated trend for the social system that is closed to new sources of energy or knowledge. One runs into the familiar Malthusian problem. At some point the expanding population runs up against the resource limits of population support and aggregate growth slows down and stops. If the resource base is reproducible or inex-

11

haustible, a static state equilibrium results with a constant thermo-dynamic throughput and a constant population size. If it is exhaust-ible, at some point population growth gives way to population decline. Thus, this mode of explanation of growth can support confident pre-diction only at times of population-resource disequilibrium—and then only for the limited historical period of adjustment.

One might suppose that a second mode of explanation could be directed to an expansion of the resource base through migration or spatial extension, but this is subsumed in the first explanation. As long as one is talking about a homogeneous resource base, this is merely a part of the process by means of which a population reaches its final resource constraints. If one is suggesting that different kinds of resources may exist in adjacent domains that can be exploited to support a continuously expanding population, then one transcends the realm of the simple growth model. The exploitation of these new resources implies that the system is open to new sources of knowledge and the new forms of behavior that their application implies. One has entered the realm of the development model that we are not yet prepared to discuss.

The second mode of explanation of the historical growth trend takes into account that human populations have developed techniques that amplify their capacity to exploit the resource base. They have devel-oped complex machines, and the more simple machines known as tools, that expand the productivity of a given resource environment. It fol-lows that the population size that a resource base can support is re-lated to the size of the stock of these human artifacts—the stocks of human and physical capital. Thus, the size of the population can be extended by increasing the productivity of the resource base through net additions to the capital stock. The size of the population then becomes a function of the size of the capital stock. But this does not really remove the dilemma encountered in the first explanation. It simply displaces the Malthusian margin. As the size of the capital stock increases relative to the resource base, the marginal productivity of the increment of physical capital declines until the resources absorbed in its generation are as great as the resulting expansion in resource productivity.

In short, we are faced with the same kind of situation as before. The only difference is that the resource-population equilibrium identified in the first case has become a resource-population-capital stock equilibrium in the second. Again, this mode of explanation of growth can support confident prediction only for the limited historical periods of adjustment marked by disequilibrium. The economic principle of diminishing marginal productivity to increasing factor proportions sees to this. In the absence of any operative welfare criteria, the size of the population comes to rest at the subsistence margin. It is no wonder that economic concepts couched in terms of simple growth earned the pejorative "dismal science."

The picture is not altered in essence if the dismal aspect is removed by transmuting the empirical form of the process through the interjection of human welfare criteria. One may introduce into this simple growth model a performance criteria derived from human goals. Thus, economists come to be less concerned with maximum size populations and more concerned with optimum size populations where the per capita dividend from the resource base either is as large as possible or meets some arbitrarily defined level. Output per capita and output per man-hour become the economic watchwords.

Translation of the process into these terms does not change its nature as an explanation of the growth in population size. All it does is redefine the margins of equilibrium between population size, resource base, and capital stocks. Changes in population size would still be associated with the historical periods of adjustment marked by disequilibrium.

One thing injection of the welfare criterion into the process does accomplish is to open the door for the first time to an active role for human prediction and planning. As long as the growth process is seen as inexorable, whether terminating or not, there is room for prophecy and resignation, but not prediction and planning in the sense that occupies man in the social process. Once man sees growth in terms of welfare-related levels of productivity he has a basis for human planning and intervention in the interest of substituting population optimization for population maximization. It is out of this that the great interest in population control and capital planning arises.

This brings us to a third mode of explanation of a historical growth trend. This explanation sees the equilibrium margin (whether expressed in terms of maximum or optimum populations) as being continuously displaced through the effect of a stream of new knowledge and technique that amplifies human powers and resource productivity and that expands the scope of the available resource environment. But this method of explaining growth requires that we open the system to new knowledge and new modes of behavior. We find we have moved once more into the realm of the development model yet to be considered. We have also demonstrated more adequately the assertion made in the previous section that in the end growth or changes in scale must rest upon true development.[2]

Allometric Growth Models. These simple growth models can be further elaborated by taking into account the fact that social systems are not simple homogeneous population aggregates, but consist of an interrelated set of functionally specialized subsystem populations. A change in the scale of the total system implies associated changes in the scale of the subsystems. To the extent that one's welfare stakes or social management responsibilities are associated with a subsystem group there is a special interest in being able to identify the relationship of its scale to total system scale.

The general principle of allometry deals with this kind of relationship. Probably the best known manifestation is associated with biological ontogeny—the process whereby a biological organism goes through constant increases in size from conception to biological maturity. Associated with this growth of the organism are related changes in the

---

[2] This forces the realization that, in the long run, the key factor is the generation of new ideas and behavioral designs to be embodied in physical and human capital. It is the amplification of human behavior inherent in the nature of the idea and behavioral design that is critical rather than the size of the capital stock. This fact is underscored by empirical research addressed to testing the crude capital formation hypothesis. For example, Abramovitz (1956) found that only about 6 percent of the increases in productivity governing the major phases of American industrialization could be accounted for by increases in the size of the capital stocks. Clearly, the increases in productivity over this period were not so much a result of the increases in size of the capital stock as they were a result of the fact that the existing stocks were being replaced with stocks that possessed a greater capacity for behavioral amplification.

sizes of component organs. There are other well known applications of this principle as well. (See Narall and Bertalanffy, 1956.)

The idea behind the allometric principle is that there are deterministic growth gradients that relate the size of the organ and the size of the organism. Different organs are commonly characterized by different growth gradients—that is, the coefficients that relate organ growth to organism growth are different for different organs. They may also be negative as well as positive. Some organs may decline in size and atrophy as the organism grows. Accordingly, a cross-section of the functional structure of the organism will appear different at different steps in the organism's growth to maturity. Thus, not only the size but the mix of the established component organs will change in a way that is deterministically related to the size or scale of the organism.

In a similar way the size or activity scale of social subsystems, be they region, enterprise, or community, is related to the size of the larger social system of which they form a functional component. Those involved in the management of both systems and subsystems have become cognizant of these relationships. It is a very common technique in business forecasting, for example, for an enterprise to begin with a projection of the scale of the national economy and apply to it some empirically derived coefficient to establish a demand estimate that can be used as a basis for business planning. In a similar way the fiscal directors of cities, states, etc., derive estimates of revenue to serve as a basis for governmental planning.

Viewing the same phenomena at a different level of aggregation, we can see the allometric principle implied in recent work of urban economists and "central place" theorists. They conclude that a large portion of the activities within an urban region is devoted to serving the population of the region. They identify differences in the scale and mix of these activities with different size urban regions. They conclude that, as an urban region grows in size, the size and mix of these "central place" activities will change in a deterministic fashion. (See Berry and Pred, 1961.) The prediction of the changes in urban size provide, therefore, a basis for urban planning through the application of the empirically established dimensions of the allometric principle.

We need to remind ourselves here that the introduction of the allometric relationship between the scale factors of systems and subsystems

15

does not alter the conclusion established in the previous section. As long as we consider these systems closed to exogenous sources of growth, prediction is restricted to anticipating the time path implied in the restoration of a total system equilibrium, and planning is restricted to bringing subsystem behavior into consonance with the allometric relationship implied by current technology.

The System Open to Resource Exchange. It is logical for us to consider, next, a growth model where the system is open to external sources of growth. If it is open to external sources of new knowledge and technique, this can serve as an engine of growth, of course. But, again, this takes us into the realm of development models. However, the typical social system is also open to the exchange of physical resources and resource-based commodities. This is an essential counterpart to the functional specialization that accompanies a social technology.

Regions, enterprises, nations—any functional social grouping—characteristically generate outputs that are produced in the service of external social groups and are not addressed directly to a self-maintenance based upon endogenous resources. This gives them a claim, in turn, upon exogenous resources. This reflects the fact that different social groups have direct access to different kinds of resource bases. Indeed, we are familiar with the fact that some among these have cut themselves off from direct resource contact and exist as intermediate activities of various kinds (e.g., the bulk of conventional manufacturing enterprises).

A consequence of this is the fact that social groupings like regions and nations often rest upon a resource base characterized by surpluses or deficits of some resources relative to the needs defined by the behavioral modes of the endogenous group. Thus, a closed system is limited in its growth to that scale supportable by the resource that forms the most constrained bottleneck. If it can act as a system open to resource exchange, however, it can trade on its resource advantages to gain external relief from its resource constraints. The availability of an external market for resources with a low endogenous marginal productivity permits the system to grow to a scale that exceeds that available to purely endogenous behavior.

16

This model is familiar to all those who read economic literature. The very popular economic base theory takes this form. The idea behind this concept is that the growth of a region or nation is dependent upon the growth of its export base. It is the exploitation through trade of its low productivity resources that provides access to those high productivity resources that open the way to larger-scale activity.

The model concept may remain in this simple form or it may be given greater system complexity in two ways. First, it may be combined with the allometric growth model to take into account the way in which the effect of the growth of the export base and its associated aggregate growth is reflected in the scale changes of the subsystem activities within the region or nation. In some cases the attempt is made to identify the coefficients that relate scale changes in the export base to a series of subsystem activities. In other cases studies go no further than attempting to evaluate the combined effect of these subsystem allometric gradients to establish a sort of aggregate regional multiplier.

An even more complex form of the model is formed if the concept is linked with an open system form of the input-output model or the classical general equilibrium model to specify the endogenous subsystem adaptations that must accompany a change in the external requirements for subsystem outputs. This form recognizes that the coefficients that specify the growth gradients linking subsystem changes in scale with the scale changes of the total system (and/or export sectors) are the product of an nth-round series of mutually interdependent endogenous subsystem adaptations. This form of the model does not leave the allometric coefficients to be empirically estimated from aggregated historical data, but computes them. In the input-output variant of the model these are computed by using fixed, empirically derived input-output coefficients that estimate the extent to which the input requirements of one subsystem modify the output performance of a functionally linked subsystem. In the general equilibrium system model the input-output coefficients are themselves variable relationships that are determined by a set of functions defining interindustry demand and supply elasticities.

In this class of models prediction and planning take on a different form. Prediction is directed to anticipating changes in the external parameters of the system. Planning is addressed to the efficient coordination or specification of the subsystem interrelationships.

It should be made clear, however, that this class of model does not provide an explanation of extended historical growth or a basis for its prediction any more than those already examined. It is true that the existence of opportunities for resource exchange may make available to the system an opportunity to grow to a larger scale, but this kind of displacement reaches an ultimate limit just as that resulting from the introduction of the role of capital stocks, above. Even if the external demand for output of the system's export sectors were to increase indefinitely, the system would reach expansion limits imposed by its resource capacities at some point.

Quite apart from this, the effect of this model concept is to place the burden of explanation and prediction upon the behavior of linked exogenous systems. But those systems must, themselves, encounter resource limits to their expansion and to the expansion of their demand for resource exchange with linked systems. One can extend the model concept to allow the exogenous systems to experience real development or behavior change, but, once again, we enter the realm of the development model.

None of the three classes of growth models discussed in this section permits us to go beyond this limitation. Growth models are clearly restricted to explaining the changes in the scale of system and subsystem dimensions or the levels of performance associated with a movement toward terminal equilibrium. Once the system is in an equilibrium position it has nothing more to say except to define the forces that maintain the equilibrium.

THE DEVELOPMENT MODEL AS PROGRAMMED LEARNING

Useful as they may prove to be in special circumstances, these growth models do not provide models of social change adequate to the understanding of its present reality. Nor do they support forms of prediction and planning relevant to the historical process we experience. Consequently, the students of economic and social change have been moving on to a different class of models. They have sought to develop model concepts that can accommodate changes in the modes of behavior exhibited by the system. They have moved into the realm of developmental models. We turn to their explicit consideration in this section.

Machine Systems versus Learning Systems. Before we proceed, we need to make clear a distinction that is important in developing the concept of the development model.

All the growth models considered correspond to a model type that can be classed as a machine system. A machine system is characterized by a fixed behavioral program. It has at its command a fixed and finite range of behavioral options (i.e., a fixed process technology); it has a specified set of criteria and techniques for interpreting environmental signals; and it has environmental sensors and signal channels of fixed capacity. It has no capacity for modifying its own behavioral program. Its behavior is fixed by its design.

This does not imply that its behavior cannot be complex or that it does not display an adaptive capacity. The most general of all machine systems, the modern computer, illustrates this. It has the capacity to receive symbolic input signals of extremely large volume and varied nature. It can subject them to logical operations of almost unlimited variety and sequence. Precisely how it performs in a given instance depends upon its program or set of instructions and the environmental information brought to its attention by its input devices. This program can be quite complex and can include both general and specific instructions along with selective criteria. Under the control of these criteria and instructions, the machine can adapt its response selectively to a variety of forms of symbolic input. It does so within the framework of a fixed program of options and selective criteria. It is limited by the degree of freedom in choice built into the program.[3]

In a similar way the most general of the growth models, the conventional general equilibrium system in economics, is conceived to operate as a machine system. The classical statement presents us with a picture of a complex activity network with a fixed behavioral program.

---

[3] This is not to deny that the modern computer can be programmed to exhibit at least rudimentary characteristics of a learning system or that progress may not be made in this direction in the future. Further reference to this will be made in a later section (p. 27).

19

The general equilibrium system is visualized as a set of industrial and final demand subsystems engaged in the physical transformation of the earth's material and energy resources in a complex thermodynamic process essential for the maintenance and reproduction of the life of the species hominid. Each of these subsystems characteristically displays a range of behavioral options. (Classical production and consumption functions both allow a range of scale, substitution, and process options.) The permutation of these subsystem options for the total economy can form an extremely large number of behavioral sets. Most of these would represent sets of mutually inconsistent subsystem inputs and outputs (implying total system disequilibrium) and would, therefore, be inconsistent with the efficiency goals of the total system. Only a limited number would represent mutually consistent inputs and outputs, and it is hypothesized that some one set (the efficiency optimum) would be better than all others.

This optimum equilibrium is found by a process of mutual adaptation of the subsystems. Each will choose a mode of behavior from among its behavioral options on the basis of environmental signals (e.g., prices and orders) submitted to a set of interpretive criteria (e.g., profit maximization). It is, however, the kind of adaptation that is consistent with a fixed program. Each change in environmental signals (e.g., prices and orders from the point of view of the production subsystem) brings a change in output levels which acts on the environment (e.g., prices and orders) once more. The subsystem is, thus, locked into a functional feedback network with the other subsystems that make up the total system. The total interacting network is assumed to be a deviation counteracting system that leads to a total system optimum. Once a stable system is achieved that makes the most efficient use (in terms of its total system goals) of its constituent resources (given its fixed behavioral program specifying both goals and technological options), it establishes a continuous flow, steady-state system.

The less general growth models (the input-output version of the system open to resource exchange, the allometric models, etc.) also function as machine systems. The only difference is the level of functional aggregation and system complexity at which the process is visualized to take place. As has been pointed out, these models offer an explanation of social change only to the extent that it can be interpreted as a movement to a terminal equilibrium state.

Set over against this concept of the machine system is the concept of the learning system. A learning system contrasts with a machine system because it has the capacity (a) *to be reprogrammed* in its behavior, or (b) *to reprogram itself* through the action of internal sources of new behavioral ideas, transformation motives, and transformation behavior.[4] In either case it comes to embody new modes of behavior or new kinds of activities and, hence, new organizational structures.

The importance of understanding the distinction between fixed program and learning systems develops in the following way. Social organization and social systems are human systems. They have developed out of the capacity of man as a learning system. Conspicuous among the social systems are the region and nation systems that are a product of the logic and order of subsystem transfers or activity linkages. These composite social behavioral systems, since they are designed and activated by man, are also learning systems—their behavior is subject to continuous reprogramming. There are very important consequences for their form and function that stem from the fact that region and nation systems are the creations of a social learning process. Their transformation, which is the primary object of our concern, can be seen as a product of the exogenous or endogenous behavioral reprogramming to which they are subject. Even in simple, relatively homogeneous social subsystems, a significant proportion of that learning process is internal to the system.

The essential point here is that the equilibrating optimizing concepts that dominate the methodological base of the growth model are, in effect, machine system concepts. If we are to move beyond the uncomfortable restrictions they impose we must come to understand how social systems behave as learning systems.

In the present section we move to examine the developmental models that are constructed upon the concept of the "class a" learning systems referred to above (i.e., the social system open to directed behavioral reprogramming).

---

[4] Both Bertalanffy (1962a) and Ashby (1962) make a similar distinction between what they call machine systems and "self-organizing systems." The relationship of cybernetics and general systems theory to the thesis presented in this volume will be considered in greater detail later.

21

Component Learning Implicit in Growth Models. Before we consider the development models of "class a" type that find explicit articulation in the literature we might point out that there is an element of programmed learning implicit in growth models.

In the context of the growth model, as the system (region, nation, etc.) moves toward some terminal state its changes in scale are associated with changes in the activity levels of component subsystems. Some of these activity levels may change within limits without any change in the social structures that underlie them because the range of change in activity levels does not impinge upon capacity limits. Characteristically, however, some of the activities will encounter capacity limits rather soon. When this happens net capital formation must take place. Since the allometric coefficients are different for different subsystems, this leads to both a change in the activity mix and a change in the structure of the capital stocks that embody the mix of activities.

We have seen that these changes do not embody learning at the level of the total social system or its social subsystems. As systems of group behavior they do not come to incorporate any new modes of behavior not previously exhibited. However, the changes in activity levels and their supporting capital structures require changes to be made in the stocks of physical and human capital. Thus, knowledge about established modes of behavior has to be embodied in human individuals and physical capital instruments that did not previously incorporate these behavioral capacities. They have to be programmed with behavioral capacities appropriate to the new requirements of social machine system behavior. Even the process of normal replacement of these individual components requires the capacity for such programmed learning. Thus, the social system conceived as a growth model must incorporate systems of programmed education and capital formation operating to transform the behavior of individual components. This is essential if the system and its organized functional subsystems embrace the capacity for self-maintenance and the capacity to negotiate changes in scale or levels of intensity.

The Machine System Open to Information Inputs. We must go beyond this, however, to articulate a model that will serve to explain how the social system (again, enterprise, region, nation, etc.) may become changed in the *modes of behavior* it exhibits.

Development models described in the literature tend to take the form of the "class a" learning system mentioned above. In other words, changes in the behavior of a system are considered to come about as a result of a process characterized in economics as comparative statics. Conceptually it is similar in character to removing one program from the control unit of a computer and inserting another. A new program is plugged in that either modifies the behavioral rules guiding the system or its technical behavioral options. One then solves the system (which can be expressed as a series of difference equations) to determine the new pattern of behavior that will emerge after the deviation counteracting feedbacks have done their job (and on the assumption that the terminal equilibrium of the total system is independent of the path of the feedback). Change is something introduced at discrete intervals from outside the system—presumably by a hypothetical chief programmer.

One might note here that the system open to resource exchange, previously discussed, also has open communication with the exogenous systems with which it is linked. However, in this model the communications take the form of simple signals that communicate the facts of the state of the exogenous world to the system so that it can make necessary adjustments in its export and import related activities. It is marked by the transmission of orders, prices, etc.

In the model of the machine system open to information, the system is presumed to have access to knowledge of new behavioral modes and the capacity to incorporate this knowledge into new activities and their supporting physical and social structure. But essential to the nature of this model is the assumption that the *new knowledge comes into being outside the system*. It exists in the form of established technique and behavior in the exogenous world that can be borrowed for application in the endogenous world.

If the system under consideration is actually under the hierarchical control of an exogenous system (as in the case of a manufacturing plant being under the directive control of a corporate enterprise) then it comes to be literally reprogrammed by the external agent. The analogy of inserting a new program into the control unit of a computer is a

23

quite accurate characterization. If the system is not under direct over-all control of the hierarchical unit (as in the case of a region within a nation or a nation in an international community) the analogy is weakened somewhat. Even here, however, in the context of this model the new behavior is seen to arise through a combination of changes directed by external agents and a process of behavioral "mimicry." In the latter case the system perceives a technique or mode of activity in practice in the exogenous world and moves to incorporate it into its own behavioral modes. It is still approximately valid to view this process as one of programmed learning.

A number of themes that are important in the development litera-ture find their place in this conceptual model. Paramount among them is the emphasis placed upon capital formation. The formation of human and physical capital is the process by means of which both new and old behavioral modes are incorporated in system behavior. One is impressed with the fact that the major portion of this literature treats education and the production of physical capital as the instruments of programmed learning. Education is designed to embody in human be-havior those skills and techniques that equip the individual to play an efficient role in the social machine. The production of physical capital is devoted to tranforming resources into those machine artifacts that play such an instrumental role in the operation of the social machine.

Another theme closely related to this model is the considerable in-terest in the diffusion of knowledge. Part of this interest is devoted to the empirical identification of diffusion rates and goes no further than an apparent search for historical generalization. But there is a consid-erable and growing interest in trying to identify the functional chan-nels through which diffusion takes place. In this connection an im-portant contrast between the concept of the system open to resource exchange and the system open to new behavioral knowledge is mani-fest. In the former case the focus was upon the *adaptation of the system* to the existing states of an exogenous world within the frame-work of established behavioral modes. In the latter case, the emphasis is upon the *adoption by the system* of exogenous behavioral modes. The literature on information diffusion is beginning to go beyond its early preoccupation with diffusion rates to a consideration of the activities and channels that facilitate the adoption of new techniques (e.g., Nelson, Peck, and Kalachek, 1967, and Lasuén). Since the system subject to analysis is not commonly under the directed control of exogenous systems, the concern is that the system develop sufficient understanding of the diffusion and adoption process so that the programmed learning can take place more efficiently.

The interest in and application of this model concept helps to explain the burgeoning interest in information classification and retrieval systems. In order to increase the efficiency of programmed learning we wish to improve the speed and accuracy with which we can search for and identify relevant technological or new behavioral options available in exogenous sources.

Prediction and planning take on a different aspect under the control of this concept. Prediction takes the form of the identification of new behavioral forms in the exogenous world. Planning takes the form of implemented programmed learning directed to forming the physical and human capital and the organization forms appropriate to the new behavioral modes.

Development Stage Concepts. The substantial interest in stage theories of development seems to fit in here as well. This is particularly true if one is concerned with the development of backward nations or regions. In these cases there is a vast range of novel technological and organizational behavior practiced in the advanced countries that may be adopted in the underdeveloped country. Therefore, diffusion and adoption are faced with a problem of choice, and it is thought that an important element in the problem of choice is the problem of sequence. The adoption of certain ideas and techniques should, perhaps, precede others.

As a guide to such a sequence, history is examined to identify the development sequences followed by the advanced countries. These are visualized as a template to guide the process of programmed learning designed to bring the backward system to modes of behavior more nearly approximating those of the advanced countries.

MODEL EVOLUTION AND MODEL COMBINATIONS

It is plain to see that the preceding sketch of concept models forms a natural progression from less to more system complexity, from shorter- to longer-run processes, and from reflexive to learning adaptations. The approaches to prediction and planning implicit in each of these concept models become progressively more general. This progression forms a natural evolutionary sequence.[5]

Another characteristic of this process needs to be made explicit. As we move from simple to more complex growth models, and from growth models to development models, the simpler models are not displaced from the scene by the more general models. Rather, their relevance becomes greater as they become part of a larger and more inclusive process. One can visualize a development model which embraces all of these model types in combination. It could be, for example, a process by means of which a system is absorbing new modes of behavior from an exogenous world through programmed learning following a historically revealed developmental sequence. In the process, there might be alterations in its patterns of resource or commodity exchange with subsystems whose traditional modes of behavior remained unaffected by the development. This might alter, in turn, the scale and activity levels of that subsystem. This, in turn, might set in motion certain endogenous allometric adaptions that generate a different mix of activities. In short, a disequilibrating adoption of new behavioral modes may be combined with equilibrating adaptations at subsystem levels to generate new levels and a new mix of activities and their associated forms of social structure.

Thus these model concepts can be combined to form still more complex models. Furthermore, the simple models may retain some utility

[5] Once again, no attempt is made here to write a documented history of economic thought. If one follows in detail the development of the literature, a case can be made that no linear sequence of ideas can be traced. Such a conclusion would not be strange. It is common in the development of thought that a few perceptive scholars may be expressing advanced concepts that have little currency in the contemporary literature. Furthermore, the students of development tend to work within the frame of a particular systems concept. Thus, those dealing with regional development may come to develop and utilize one concept model at a later time than those dealing with the concept of national development or the development of an industrial enterprise. Examined closely, the literature may give a confused picture of the sequence of the emergence of such concepts. But if one backs away and traces the trail that appears to have been taken by the dominant concerns of those engaged in prediction and planning, some such sequence is observable.

and validity as a guide to action in cases where system complexity is limited and predictive goals circumscribed. We shall return to this theme in chapter IV.

DEVELOPMENT MODELS AS CREATIVE LEARNING

For the present let us utilize this sketch to identify the need for a still more general approach to prediction and planning. All the concept models in the sketch share a common defect. They are deterministic in the sense that the terminal state of the system is predetermined once the process of change inherent in the model is initiated. All of them push off into the exogenous world a consideration of the process by means of which the external parameters become changed or the new modes of behavior come into being. Let us examine this theme more carefully.

There seems to be a great deal of reluctance to incorporate nondeterministic processes in the concept models of structure and change of social systems. This may stem in part from the philosophical presuppositions of some social scientists. Undoubtedly it also stems from uncertainty about how to construct a meaningful nondeterministic model and a reluctance to deal with the complexity such a process might be presumed to embrace. Consequently, every effort seems to be exercised to stay within the deterministic frame.

The attempt to do so may lead one to try to construct a model of a learning machine or self-organizing system. In the cybernetics literature Ashby (1962) characterizes such a model as a "machine with input." This is a formalized counterpart to our concept of the machine system open to information.

Ashby begins by asserting that no machine can be self-organizing in the sense that it can improve its own organization or performance. He then confronts the apparent contradiction presented by the computer that writes its own program. He points out (pp. 266–69) that this comes about only in a "machine with input" or a hierarchical machine system: ". . . the appearance of being 'self-organizing' can be given by the machine . . . being coupled to another machine. . . . Then the part . . . can be 'self-organizing' within the whole. . . . Only in this partial and strictly qualified sense can it be understood that the system is 'self-organizing' without being self-contradictory."

This means that a machine, say computer A, may be incorporated into a larger machine system by coupling it with another machine—computer B. Computer A may be programmed to solve a range of problems presented to it and it may have, accordingly, a range of preprogrammed adaptive behavior. Computer B may be programmed with an operational capacity that will enable it to modify computer A's program or behavioral capacity under certain circumstances. Thus, if computer A is confronted with a problem it is not programmed to solve, computer B might come into action to modify A's program in a way that will enlarge or modify its operating capacity. In order for B to accomplish this, however, it will, itself, need to be preprogrammed with a set of criteria for identifying computer A's performance as inefficient or inadequate, coupled with a set of operating processes that (together with the criteria) can generate a change in A's behavioral capacity.

In this way machine systems can display quite complex adaptive behavior and take on the appearance of learning systems. However, such a complex machine system can act like a learning system only to the extent that it can be preprogrammed with selective criteria and well-defined learning processes. The control machine has to be given this capacity by another machine or human being. If it is another machine, it, in turn, must be similarly instructed. Thus, if the learning capacity of such a complex machine system is to be further generalized it will have to be governed by a hierarchical series of control units embodying successively more general criteria and process instructions. But this turns into an infinite regress that can only be terminated with a "god" machine that incorporates all wisdom and knowledge and builds into the machine system a preadaptive capacity for every possible contingency.

The end result is the creation of a hierarchy of programmed learning. In this vein, one could visualize the exogenous world as having the capacity to reprogram the system under examination (enterprise, region, nation, etc.) by imparting to it its own previously acquired behavioral capacity. Sooner or later, of course, there would be no new knowledge or no new modes of behavior left to transfer and all social change would stop—unless, that is, the exogenous world were conceived to have its own exogenous world that was acting to reprogram it.

It is readily apparent that such an approach does not get to the root of the fundamental processes of social change. Even if we do not understand these processes well, casual observation and limited insight make plain that the process is, at base, more open-ended and creative in form. Such an infinite regress of programmed learning does not resolve the question of how new modes of behavior arise in the exogenous world. Furthermore, we have to recognize that programmed learning of the kind implied in the development model discussed above is not adequate to explain the behavioral changes in the endogenous system either. There are cases, of course, when an external agent directs behavioral reprogramming in a system. A corporate enterprise, for example, may reprogram the performance characteristics of one of its manufacturing plants. Even regions and nations may occasionally experience a change in some aspects of their internal behavior brought about by the direct intervention of an external agent.

We have to bear in mind, however, that these social systems may be characterized as somewhat like ecological coalitions of subgroups. They are only occasionally completely under the control of a directing center of authority. The very boundary between what is considered endogenous and what is considered exogenous is arbitrary and arises in the context of the exercise of prediction and planning in the interest of solving social problems. The diffusion of new knowledge into a system from outside sources often requires the initiative of the system itself, or of some system component.

It requires a conscious search of the exogenous world for new modes of behavior to solve endogenous problems or achieve endogenous goals. In short, new behavior is instigated by the system's own purposes. Even so, one might still hold the image of deterministic programmed learning by assuming that this outreach is a basic part of the behavioral characteristics of the social system (although an assumption that empirical evidence tells us so is certainly not always fulfilled). This kind of supposition leads to the concern with diffusion rates and ways in which this outreach can be made more efficient—in short, ways in which the deterministic process can be recognized and, therefore, made more efficient in operation. It also leads to the attempt to visualize this as

29

the manifestation of some deterministic historical process embracing developmental stages.

But this is not the end of the problem. If the process were to remain deterministic, the techniques and new modes of behavior that preexist in the exogenous world would have to suit precisely the functional and organizational requirements endogenous to the system. This rarely happens. Even when borrowing new behavioral modes a social system characteristically has to modify them to fit the functional and organizational peculiarities of the system. In the interest of realizing its goals it may also find no appropriate preexistent behavioral modes and may invent a new mode of behavior independent of the influence of exogenous systems.

In short, an important part of social change—indeed the most important part—is the product of an open process of creative learning. In contrast to the development model as programmed learning, we are faced with the development model as creative learning. Such a system has the *capacity to reprogram its own behavior*. It has a learning capacity that is endogenous in character and does not depend upon external agents. Since this learning capacity rests upon the unique learning capacity of human agents, no social system is too small to contain *within itself* some capacity for open-ended learning. This is the "type b" learning referred to on page 21.

If we are going to reach once more for a model concept that provides a more general approach to prediction and planning, it is in this direction we must reach. We must inquire into the processes by means of which creative endogenous social learning takes place.

Biological Evolution as a Learning System Analogue

Faced with such a task and purpose one naturally looks about to see if there are areas of scientific study where the operation of nondeterministic learning systems has been made the object of explicit attention. There does exist such a body of study in the form of the scientific literature devoted to the modern synthetic theory of biological evolution. We turn to this body of study in the beginning of this effort. The process of biological evolution seems to be a fundamental point of departure because, as a learning system, it operated prior in time to social systems and, indeed, human learning capacities and social evolution are products of the life system. It may provide us with a useful reference point for entering the domain of social learning.

At an earlier stage the life sciences went through the same kind of painful reexamination and generalization of basic concepts that economics and the social sciences now seem to face. It is instructive to review the evaluation of this conceptual problem made nearly fifty years ago by Alfred J. Lotka, one of the great pioneers in modern biology. (Lotka, 1924, chaps. 21–22.)

It was the tradition of biology in an earlier phase to borrow system concepts from the relatively simple systems of physical chemistry. At that time "it *was* common custom . . . to assume that a sufficiently slow change of one parameter defining the state of the system brings in its train a succession of states each of which is essentially equilibrium." This assumption was based upon the principle of continuity. There was another tradition following the generalized principle of Le Chatelier in chemistry. This principle stated that a system change could be analyzed by looking at the beginning and terminal states. It "inquires only into the ultimate effect, upon equilibrium, of a given total change in a parameter, leaving aside all questions relating to the path by which the displacement of equilibrium takes place."

Both of these cases are attractive because they yield with comparative ease to analytical treatment. The economist will have no trouble in identifying these concepts with common custom in economics. The principle of Le Chatelier is readily identified as the principle of comparative statics in economics. The economist, as the biologist before him, is tempted into frequent application of these principles in analysis.[6]

Lotka pointed out, however, that the legitimate application of these principles in biology rests upon two conditions being fulfilled. First, the system must be a stable system. Second, the displacement must be independent of the path of change.

---

[6] It is of interest in this connection that Paul A. Samuelson offered a proof of the principle of Le Chatelier in his *Foundations of Economic Analysis* and employed it as an integral part of his analytical framework. The introduction to his paperback edition (New York: Atheneum, 1965) also credits Lotka as an important source of inspiration for that work.

31

Lotka goes on to identify characteristics of biological organisms, communicable disease models, and biological evolution where one or both of these conditions are not fulfilled, concluding that many phenomena of importance to the life sciences do not fall in this class. In the intervening years life scientists have devoted considerable effort to the study of these open transformation systems, and out of their combined efforts a modern synthetic theory of biological evolution has emerged—one more complete, internally consistent, and analytically powerful than any earlier set of concepts. The development of these concepts in modern biology tempts one to seek there some guidance in thinking about the dynamically open transformation processes we face in the social sciences.[7]

One is tempted because it seems evident that one or both of the Le Chatelier conditions are not fulfilled by many aspects of regional and national economic development. We continue to force these phenomena into the old conceptual mold. Indeed, the economic development problem is commonly characterized as an arrested level of behavior that needs to be deliberately reprogrammed. It is difficult for a discipline trained to analyze equilibrating mechanisms and inclined to view equilibrium departures as pathological to adjust to the consideration of a stable equilibrium as pathological and to deliberately specify induced disequilibrium as the prescription. It becomes even harder to specify the mechanism or stages by which the transformation is to be effected. Even when this change in perspective is attempted, the reform is often partial. There is a tendency, noted above, to consider more primitive economies as "suboptimal" and to view the process of change as a "displacement" to an activity complex more nearly matching the standards of advanced Western economies. Very often the terminal state of the displacement is assumed to be independent of the path.

Increasingly the big questions in the development field are: How

---

[7] Several excellent treatments of the modern synthetic theory are available. Among the most significant are those by Huxley (1964), Mayr (1963), Rensch (1959), and Simpson (1949, 1953). Dobzhansky (1962) applies the concepts of synthetic biology with special reference to certain aspects of human evolution. The author is greatly indebted to all these works but he has drawn especially heavily upon Rensch and Simpson. Chapter One in Mayr is particularly good in gaining some conception of the development and definition of modern synthetic biology.

can we stimulate disequilibrium? How can it be directed in a constructive rather than a pathological sequence? How can the sequence be
cumulatively sustained? The restive spirits are being heard more frequently, but such questions are as often implied as made specific, and
they do not as yet constitute an organized systematic inquiry.

In short, we need to see the process of economic change as incorporating something more than equilibrating adaptations or deterministic
historical trends, and inquire more explicitly into the more general
processes characteristic of learning systems, seeking more explicit criteria for distinguishing progressive from regressive behavioral change.
A review of modern synthetic biology may provide us with a study
guide for entering this domain.

Risks Inherent in Analogy

At the outset we should recognize the risks that are involved in the
use of biological analogies. In the past, similar attempts have often
fared poorly.

In an earlier phase of the social sciences there was a great hope that
the analogy drawn from biological evolution might serve to aid the
understanding of the structure of social systems and their transformation. Few of the social science disciplines remained unaffected. The
early sociologists, like Spencer, became biologizing sociologists. (See
Greene, 1959.) The economist seized upon different aspects of the biological analogy so that social Darwinism generated both the Marxian
stages of development and the idea of laissez-faire. Cultural anthropology more or less began on an evolutionary "kick" and such early
writers as Tyler and Morgan were thorough-going evolutionists. (See
Kroeber, 1960.)

Then the reaction set in and the concept of evolution in the social
sciences came to be thoroughly repudiated. Such outstanding social
scientists as Franz Boas in anthropology, William James in psychology,
and Joseph Schumpeter in economics reacted against the early mis-
reading of evolutionary theory that led to the notion of unilinear
progress and indeed against hidden value presuppositions that reside
in the term "progress." "Cultural relativism" became the order of the

day in anthropology and "biological analogy" became a pejorative term throughout the social sciences. (See Sahlins and Service, 1960.)

It now seems clear that the reaction was excessive. We can also perceive the defects in the earlier application of analogy that led to its rejection. One limitation is inherent in the learning process itself. At the time when biological analogies were popular in the social sciences biological thought was still incomplete and inconsistent in many respects. The modern synthetic theory of evolution provides us with a much stronger base for analysis than formerly existed.

A more important defect, however, was the use of reductionistic logic. Just as in the earlier phases of biological study there was a tendency to reduce biological systems to the machine systems of physics and engineering, so during this phase of the social sciences there was a tendency to reduce social systems to an identity with the components of biological systems. In the application of the biological analogy most of the energy was absorbed in the search for direct counterparts to biological processes in the social process. This yielded some striking insights, but ended in freezing social concepts through the identification of social process with a process model insufficient to describe social reality.

Yet it seems probable that the biological analogy has an appropriate use that can avoid this risk. The evolution of the characteristics of biological systems that are exclusive to the biological process did not make mechanical processes irrelevant in biological study. Under certain circumstances some biological subsystems of organisms, the organisms themselves, and, indeed, certain transformations of biological populations act like machine systems and can be usefully analyzed as such. This does not deny the fact that they retain at the same time the potentiality for behavior uniquely different from machine systems. Similarly, under certain circumstances social systems may act like machine systems or biological systems and still retain the potential for behaving in ways unique to social systems. Thus, it is appropriate to look for and identify process similarities between biological systems and social systems and to learn from their suggestive content.

Beyond this, the use of such an analogy can be more rewarding if it is undertaken primarily as an aid in highlighting the unique characteristics of social systems. Social evolution may stand more clearly revealed when seen against the background of its antecedent historical processes.

Today it seems unscientific to deny the social sciences such insights as may be gained from this great body of scientific work. Up to this point some of the leading modern biologists (Dobzhansky, Simpson, Rensch, Huxley) appear to be ahead of most social scientists in specifying some of the unique characteristics of social processes—albeit they do not elaborate this theme. Used properly, therefore, the modern synthetic theory of biological evolution can be a useful and justifiable device for entering the domain of learning systems.

Objectives

Under the impulse of the motivation provided by the growing importance of more general predictive goals in social research and public policy, and guided by the conviction that this quest leads us to an explicit consideration of the nature of learning systems and a greater understanding of the process of social learning, we set out upon the following intellectual reconnaissance:

We shall first review the principal characteristics of the modern synthetic theory of evolution and establish the sense in which it functions as a learning system. Next we shall undertake to identify the principal characteristics of social evolution. In doing so we shall attempt to make plain that the process heritage of its antecedent, biological process is carried forward but in transmuted form. We shall also attempt to establish the ways in which the biological process falls short in describing social processes and to identify some of the unique characteristics of social development. In this fashion we hope to develop a preliminary understanding of the components of social learning.

At this juncture we shall return once more to our entry portal and seek to understand how these concepts extend one's framework for dealing with the problems of regional and national development. We will close with an examination of the more general implications of social learning concepts for economic and social science research.

In undertaking such a project the author is exposing himself to an additional risk not already identified—the risk of dilettantism. Even though this effort is termed a reconnaissance, the author cannot avoid trespassing upon fields of knowledge of which his mastery is limited.

35

Introduction

Doubtless some of the points elucidated here will not satisfy the experts. His conscience is troubled on this account, but it is bolstered by the conviction that there may well be a valuable function in this kind of error.

Part II

EVOLUTION AND LEARNING SYSTEMS

II   Biological Evolution

THE TASK OF DESCRIBING, in brief compass, the principal features of modern synthetic biology is a formidable one. Let us begin by describing the mechanism of biological transformation and then consider the large-scale features of that process. In interpreting the ways in which it performs as a true learning system, we shall learn that one of the most significant large-scale characteristics of biological evolution is the gradual evolution and transformation of the learning process until the process of biological evolution generates the process of social evolution as an outgrowth of itself.

Self-Reproduction and Self-Maintenance

The essential characteristics of the biological process are the reproduction of self through the genetic process and the self-maintenance of the resulting organism to a stage permitting self-reproduction once again. These twin characteristics are the sine qua non of life. Evolution has accomplished this basic step by providing for the replication of genetic material by a variety of mechanisms. Each new organism (phenotype) is set on its energy-seeking, energy-utilizing path by an information relay whereby it receives from an earlier contestant a set of instructions setting the pattern and limits of its life cycle (its onto-genetic development), and passes them on, in turn, to the next runner in the race.

39

The Process of Biological Transformation: Phylogenesis

But life does more than reproduce itself. It transforms itself. It engages in a continuous, restless, opportunistic search for new forms and new functions. Each new experiment in self-transformation is tested in life's crucible to determine its relevance to the reality of the natural environment experienced by the organism. The maladaptive innovations are weeded out because their carriers cannot survive. The adaptive innovations are maintained in the species through genetic replication. They strengthen its members for the next stage of the contest.

The mechanism of this process of transformation is now well understood. It is a complex function of changes in the genetic material of the organism (genotype) and changes in the environment that is its host.

The change in genetic material comes about primarily through mutation and genetic recombination. These alterations of the inherited characteristics constitute the fountainhead of evolution. It appears that spontaneous mutations can happen to almost every gene and can, therefore, alter almost every possible form and process (morphology and physiology) of living matter. As yet the process of gene mutation is imperfectly understood, though geneticists are commencing to make rapid progress through the analysis of bacteria and virus mutations. It is believed that they are fully random or nondirected in nature and that they can cover the full range of heritable characters.

The changes thus wrought in the genetic material of an individual member of any biological species can be infinite in variety. That the life forms and processes resulting are not infinite is due to the fact that each mutation or genetic recombination must survive two tests: (1) The "new idea" must be consistent with the continued viability of the organic system in which it arises. It cannot violate the "logic" of the system in such a way as to foreclose altogether the development of the individual (the ontogeny of the phenotype). It is an interesting aspect of this process that the vast majority of sharply distinct mutants or recombinations proves to be lethal. (2) The nonlethal mutations must, in turn, meet the test of relevance. If the new form or function aids the developing organism in seeking and utilizing energy in its host environment, it strengthens its survival characteristics. If it is maladap-

tive, it weakens them. In an interbreeding population of such organisms the statistical distribution of race or species characteristics in the
gene pool will become progressively modified under the influence of
this form of natural selection bringing about a gradual change in the
characteristics of the species. The favorable genetic combinations
spread and the unfavorable ones disappear. In the process, phylogenetic
change takes place.

This selection process is aided and abetted by the fact that changes
in the environment also take place. Under the effect of such changes,
formerly adaptive characteristics become maladaptive and disappear
from the gene pool. Certain preadaptive characteristics or latent genetic ideas (existing in specific genotypes that never became widespread
in the gene pool because of their former irrelevance) now commence
to spread. The organizational (morphological) and functional (physiological) range of the new mutants that can be viable is altered.

The speed and extent of the changes depend upon the process of
selection rather than upon the variability of the material produced
by mutation. The various agencies that maintain selective pressure
upon the gene pools of the biosphere are quite diversified. They include such factors as changes in temperature, moisture and radiation;
the presence of predators and changes in their efficiency; the effect of
parasites and infectious diseases; and the competition for food, territory and nesting sites. Furthermore, mobile organisms may migrate to
different environments under the pressure of deficient food sources.
Many of these factors have the effect of generating fluctuations in
population size which, in turn, have a tendency to accelerate and amplify the selective effect.

Since the evolutionary effect is in every instance the combined effect
of mutation rates and the variety of selective factors, it is plain that
there can be variations in the transformation paths followed. It is now
the most widely held view in the life sciences that all known organizational and functional transformations of biological organisms and their
species populations can be explained by the operation of these genetic
and environmental factors. This point will be enlarged upon later in
the chapter.

The operation of this mechanism in bringing about the gradual
transformation of the genetic material in the biosphere is the process

known to life scientists as phylogenesis. Simpson (1953, pp. 377–78) characterizes it in the following way: "The history of life as it has existed in nature is a vast succession [phylogeny] of the ontogenies [life cycles] of organisms, all of the ontogenies being connected by the fact that each arises with material continuity from one or from two others.... Looking more closely into the pattern we see that it involves also the organic changes that have occurred in the sequence and the rates at which these changes occurred.... Thus the pattern of phylogeny brings together everything in the complex, integrated phenomenon of evolution."

The phylogenetic process splits into two major subprocesses recognized by the modern life scientist. The source of distinction has to do with the differences in the nature of the adaptation undertaken by biological populations under different circumstances. The first of these we shall call adaptive specialization and the second, adaptive generalization.[1]

In the most general terms, adaptive specialization generates the incremental refinements in form and function of a biological population with respect to its environmental niche. Adaptive generalization generates evolutionary novelties—i.e., larger changes in form and function that have the effect of redefining and broadening the adaptive zone relevant to the population. We turn our attention to enlarging the content of these distinctions.

---

[1] This is a terminology not common in the biological literature. There, the former usually would be identified as phylogenetic branching or adaptive radiation and the latter as progressive evolution. This less conventional terminology is used here because it better suits our narrative requirements.

Other similar terminological distinctions in the literature are "specific" versus "general" evolution (Sahlins and Service, 1960), "splitting" versus "quantum" evolution (Simpson, 1953), and "cladogenesis" versus "anagenesis" (Rensch, 1959). The new terminology is adopted here for three reasons: (1) it seems more descriptive of the major characteristics of the two kinds of adaptation; (2) it helps us to reinterpret biological evolution as a learning system; and (3) it avoids possible normative or orthogenetic implications of such terms as "progressive" evolution. In this brief summary it seems unnecessary to interpret or redefine traditional terms that have descended from an era when they had a less precise scientific content than their present usage allows; nor does it seem desirable to make the fine distinctions between the differences in usage that exist in the biological literature.

Adaptive Specialization: Phylogenetic Branching

By long odds the largest number of adaptations follows the mode of adaptive specialization. As a common interbreeding biological population finds its adaptive zone characterized by some environmental variation (forming subzones which exhibit differentiating as well as common environmental characteristics), this process of adaptive specialization usually leads to subsets of the interbreeding population (characterized as demes, races, or subspecies by the biologist) for which the mean characteristics of the genetic pool of the subset differ from the mean of the total population. The organisms in these subsets then exhibit some differences in structure and behavior from the total population.

Often the subsets become progressively isolated under the influence of these adaptations until they no longer retain the capacity for interbreeding with the parent population. When this happens a new species is formed. The overall effect of this attribute of adaptive specialization is phylogenetic branching or adaptive radiation. A common interbreeding gene pool becomes progressively differentiated and separated into distinct subsets as the members of the biotype radiate into a variety of environmental niches or adaptive zones and as the genetic transformations take different paths under the influence of different challenges of the environment.

Consider these examples of the incremental specializing character of the adaptations. First, climate gradients lead to distinct changes in form and function differentiating bird species. As the climate becomes progressively cooler one can witness an increase in body size, a shortening of the relative length of extremities (tail, bill, limbs), more pointed wings, an increase in the number of eggs per clutch, and an increase in the migrating instinct. All are adaptations favoring survival in cold climates. (See Rensch, 1959, p. 43.)

Another example familiar to even amateur bird and butterfly watchers is offered by melanic mutations. These are adaptations in protective coloring that take place without major structural or behavioral changes. Some of these, quite recent and referred to as industrial melanism, occur where certain types of moths in environments subject to industrial encroachment have found in sooty surroundings a selective advantage in a black mutant that becomes spread through the population in successive generations. (See Rensch, 1959, p. 20.)

How these adaptive specializations can lead to progressive differentiation and branching of parent populations is shown by one family of songbirds. This family, stemming from a common historic population, has differentiated into 153 species, each of which contains 2 to 31 identifiable races or subspecies which in turn retain some overlapping of genetic pools through limited interbreeding. (See Rensch, 1959, chap. 3.)

These radiations of biotic types do not take place at a persistent or uniform rate through time. Explosive radiation is apt to take place during periods of major change in environment or following a particularly fruitful biological innovation (such as the wing or controlled internal body temperature) that opens to invasion a new and broader adaptive zone. In the former case new habitats may be created by the formation of new deserts, new mountain and alpine zones, new lakes from craters and glaciers, and new seas in the course of oceanic invasions. In the latter case, some species may be displaced by stronger competitive types or the new innovation may allow the new biotype to share the same geographical domain with other species because they make use of different aspects of it.

Each of these radiating specializing lines of genetic transformation is characterized by improvement. It is an improvement in the sense of a gain of thermodynamic efficiency or organism viability within a given environmental range. The biotic type assumes the form and function that corresponds most closely with success within a given range of environmental characters. Following a flowering of biotic types comes a period when the variety of types supported by the environment is pretty much fulfilled and the genetic transformation takes on the task of perfection of type rather than its multiplication. Many of the adaptations during the early phases of radiation do not lead to optimal anatomical structures and may be superseded gradually by types of superior construction. During this phase phylogenetic specialization generates the closest adaptive correspondence of biotic type and environment, and the rates of genetic transformation slow down. The environmental niches fill up and each line of adaptation becomes more closely matched to environment. Thus, phylogenetic branching

44

is an equilibrating or optimizing process leading to stabilization of each phyletic line. Phylogenetic saturation of occupied adaptive zones takes place.

These forms are optimal and stable in an environmental milieu that is static-state in character, but the specializing adaptations that generate the subzone efficiencies render each type peculiarly vulnerable to a subsequent change in environment. This results from another attribute of genetic transformation. Except for minor qualifications, they are irreversible in character; adaptive radiation is terminating in nature. Thus, environmental changes often transform earlier adaptive specializations into cruel traps. As a changing environment passes beyond the range of a gene pool narrowed and made less versatile through specialization, it often forces the extinction of whole species. Just as in species formation those individual organisms fail to survive whose genetic range is inadequate to match the requirements of a changing environment, a species that generates a narrower genetic range (genetic pool of the population) through specialization may, when faced with environmental change, fail to support a dynamic adaptation and thus bring about extinction of the biotype.

Adaptive Generalization: The Path to Organism Adaptability

Several characteristics distinguish adaptive generalization from adaptive specialization. There are differences in the general characteristics of the resulting structural innovations, differences in the behavioral innovations, and differences in the relationship of the organism to the environment resulting from each mode of adaptation. Expressed in the broadest terms, these differences can be summarized as follows. Adaptive specialization characteristically generates incremental adaptations in form and function to special conditions of the environment (the fine-scale differentiating characteristics of its niche). Adaptive generalization characteristically generates new forms or functions often designated as evolutionary novelties. They result in new structural types, such as the evolution of the wing in the reptilian lineage, and they are more apt to involve harmonious changes in whole systems of organs. As we shall see later, they are subject to ex post identification but not to ex ante determination.

45

CONTRASTS IN THE ORGANISM'S ENVIRONMENTAL RESPONSE

The different relationship the resulting organism forms with its environment offers one of the most significant distinctions between the two modes of adaptation.

Adaptive specialization usually has the effect of restricting the range of the environment relative to the biotype and closing its access to a broader range or a larger adaptive zone. As a differentiating population works its way into an environmental niche, it tends to make those adaptations that are appropriate to the special features of that niche. In so doing they progressively foreclose access to ranges of more general adaptive zones that do not exhibit the distinctive features of its niche. The efficiency of specialization tends to reduce the organism's capacity for behavioral versatility.

The effect of the adaptive sequence is a progressive improvement in the correlation of organism structure and behavior with environmental constraints until a stabilizing, closing adaptation results. Stereotyped organism behavior is induced.

In contrast, adaptive generalization has the effect of opening up new ranges of the environment. The changes in form and function act to generalize structure and broaden behavioral response in a way that displaces environmental constraints and enlarges the adaptive zone available to the population. While adaptive specialization generates a cumulative improvement in behavioral adaptation, adaptive generalization brings about a cumulative improvement in behavioral adaptability on the part of the organism. Instead of tending to terminate and close, it tends to open the way for a cumulative sequence of adaptations that can permit further improvement.

A few examples will help make this distinction clear. The innovation of the wing in the reptilian lineage is an example of a generalizing structural change that opened up a whole new range of the environment to the evolving reptilian vertebrates that ended by producing the new vertebrate class we know as birds. The subsequent adaptation of protective coloration that permitted the survival of one of the resulting radiating species in a specializing niche is an adaptive specialization. The development of homoiothermy (internally regulated body temperatures) in the birds and in the mammalian lineage vastly extended the terrestrial range for their lineal descendants and constituted one of the major generalizing innovations in evolution. In contrast, the changes in body size, size of heart, and number of eggs per clutch characteristic of a bird species that permit its survival in an arctic environmental niche constitute adaptive specializations. It does not expand the environmental range of its lineal descendants.

The meaning behind this contrast in the two modes of adaptation as they relate to the environment will become clearer as we examine other differentiating characteristics.

STRUCTURAL CONTRASTS

Many of the attributes of adaptive generalization that have been identified and discussed in the biological literature, can be grouped under two broad characteristics. One type of generalizing improvement has to do with the organism's internal organization; the other has to do with its external response. The same adaptive innovation may serve to improve both. These adaptations have tended to improve the organism's behavioral plasticity or versatility.

The cumulative effect of generalizing adaptations has increased the complexity and rationalization of the structures of organisms in a way quite distinct from the effect of adaptive specialization. The former derives from the fact that adaptive generalizations have tended to generate modifications in related sets of internal organs. The behavioral adaptations that permit the development of a new set of functions like walking or crawling on land, or flying in air, or climbing in trees require complex multiple changes in organs to support the new functions. For example, walking on land requires a number of structural modifications to provide for direct air respiration, to prevent dehydration, and to provide solid tissues and sturdy organs of locomotion because the full weight of the body must be supported (in the water the weight of the body is reduced by the weight of the water displaced). Each of these adaptive generalizations requires, in turn, multiple changes in skeletal structure, gills and lungs, and epithelia. As Rensch (1959, p. 75) points out, "... sensory epithelia can function only if kept in a moist condition and the organs had to be withdrawn into the interior of the body."

47

The increasing complexity has been characterized by increased differentiation and specialization of component organs. For example, the function of visual pattern recognition or form perception requires specialized sense cells that are separated optically by specialized pigment layers and a lens to focus the light, a diaphragm to control the quantity of light, etc. Increased internal specialization and organism complexity are dictated by the requirements of working efficiency in coping with a new adaptive zone.

In turn, these increased complexities and specializations have led to increased subordination of the parts and increased centralization of the major functions—and in particular the control functions of the nervous and chemical control systems.

In contrast to the cumulative, multiple organ, complex system effects of adaptive generalization, the effects of adaptive specialization commonly are simpler and more limited in their structural changes. A melanic adaptation in the color of bird feathers, for example, results in only a superficial modification of a single structural element.

BEHAVIORAL CONTRASTS

A similar contrast is found when the two modes of adaptation are compared for their effects on behavior. This has already been touched on in the section dealing with a resulting organism's response to the environment. We observed there that adaptive generalization has the effect of broadening the environmental range of the transforming biotype and that it does so by developing greater behavioral adaptability rather than finer behavioral adaptation. Thus, adaptive generalizations have the effect of increasing the versatility and plasticity of the members of a population. This adaptability has the effect of increasing the organism's independence from or control over elements of the environment instead of increasing its dependence, as in the case of adaptive specialization.

There are three key ways in which adaptive generalization brings about adaptability or versatility.

The first is through improvements in environmental tolerance, an outcome which is often acquired when the transforming biotype internalizes a part of the environment or places it under internal control. Consider the example of homoiothermy or warm-bloodedness. In the course of adaptive generalization higher animals have developed the capacity to control internal body temperatures independent of a fairly wide range of external temperatures. There are other examples of the emergence of biochemical regulating mechanisms. There is the capacity in some biotypes to increase the number of red corpuscles in response to higher altitudes and lower oxygen supplies. Somewhat similar is the development of hormone and enzyme systems that permit ingestion of a wider variety of foods. All of these developments phenomenally extend the environmental range available to the biotype.

The second way is through improvements in mobility. Improvements in vertebrate limbs that permit migration in the effort to "choose" a more favorable environment; the greater flexibility of the neck that could bring the sensory organs of the head into closer relation to environmental stimuli; and the greater flexibility of the tongue that improved chewing efficiency—all demonstrate how improved mobility has favored organism adaptability.

The third and perhaps most important way in which adaptive generalization promotes adaptability is through improvements in the organism's discriminating faculties. The generalizing improvements in photo-receptors, or eyes, are a striking example. The growing complexity of central nervous systems is even more important. "More complicated spinal cords and brains enabled their possessors to acquire highly complicated reflexes, instincts and actions guided by control mechanisms (e.g., feedback mechanisms), by experience or even by insight. Thus, environmental stimuli were not reacted to simply by a rigid inherited behavior pattern, but by responses to each special situation." (Rensch, 1959, p. 296.)

These sources of improved adaptability are not independent. Improvements in mobility, for example, are essential to improvements in the range of perception that is essential to behavioral discrimination. But these categories serve to illustrate the distinctive mode of adaptation we characterize as adaptive generalization.

The improved behavioral adaptability of higher organisms resulting from this process finds its structural counterpart in the increased complexity and centralization of organism structure discussed in the previous section.[2]

[2] Rensch (1959, chap. 7) is especially good in describing the two broad characteristics of adaptive generalization discussed in this and the preceding section.

THE SPECIAL SIGNIFICANCE OF THE EVOLUTION OF THE NERVOUS SYSTEM

The role of the evolution of the central nervous system in leading to improvements in behavioral adaptability is especially important to our story and is worth additional emphasis.

In a sense the history of evolution can be characterized as the generation of organisms with nonrandom behavioral responses to the environment. The whole thrust has been toward improving the correlation between the environment and the behavior of organisms.

Adaptive specialization accomplishes this by reprogramming maladaptive phenotypic behavior through a selective alteration of the genotype. Specialized organisms tend to exhibit stereotyped behavior during their life-cycle development as they go through the repetitive rounds of energy and material transfers and transformations with their external environment. They have a limited capacity to modify their own behavior when burdened with maladaptive behavioral patterns. Even where some behavioral adaptability exists on the basis of previous generalizing adaptations, there are many genetically conditioned aspects of behavior not susceptible to modification by the organism itself. Hence, an improvement in the environmental relevance of organism behavior requires a modification of its genetic base. Environmental selection reprograms the genotype.

Adaptive generalization has to follow the same basic route in improving the correlation of environment and organism behavior. It is accomplished by the same phylogenetic process and, similarly, requires the modification of the genetic base of organism behavior. But it accomplishes this end by steering the process along a different route. Instead of modifying the genetic base of behavior by pruning and reshaping a stereotyped pattern, it generates a more general behavioral pattern that provides the organism with the power to act *to modify its own behavior* during its life cycle (ontogeny) in ways that improve the correlation of environment and behavior *without sole recourse to another round of selective transformation of the gene pool of the species and associated organism genotypes.* In this way adaptive generalization operates within limits to substitute organism adaptability for genetic adaptation.

It cannot be emphasized too much that this active discrimination in adaptive behavior is served best by the opening generalization—the development of the nervous system and brain. It is no accident that this is the "information system" of the organism. This is the ultimate correlating mechanism. The nervous system components are differentiated in functional terms into receptors, interpreters, and effectors. (Interpretation, in turn, requires memory and idea association.) It is engaged in (1) producing information about the environment, (2) processing it, and (3) reacting upon the environment. It is, thus, the necessary basis for the formation of versatile plastic environmental responses.

It is, at the same time, the necessary mechanism to bring about the internal rationalization of the components of the organism. The increased complexity of "higher" organisms with large numbers of differentiated components and more subsystem specialization requires some form of internal coordination. Some form of systemic unity must be imposed upon these organs in the interest of organism viability. The physicochemical mechanisms that served the more primitive organisms in this regard have too limited a channel capacity (for carrying information) to support the coordinating function imposed by increasing complexity. The nervous system becomes increasingly superimposed upon the older control mechanisms.

The nervous system and the brain serve together to coordinate internal organic functions into a total system response and coordinate total system response (organism behavior) with the supersystem that is the environment. Indeed, Teilhard de Chardin (1959, 1966) characterizes progressive evolution as progressive "cerebralization." Transformations represented by the main trunk of the phyletic tree are such that successive dominant types register higher grades of cerebral advance (characterized by both cerebral mass and cerebral complexity) until crowned by man.

Adaptive Generalization—Still Phylogenesis

At one time the evolutionary novelty that is a product of adaptive generalization was thought to be unexplainable in phylogenetic terms. Since Darwin, many biologists have attributed the differentiation of races and species entirely to adaptive specialization. More recently,

51

however, some biologists have thought that evolutionary novelties (representing major changes in organism structure and behavior) could not be sufficiently explained by undirected mutation and environmental selection. They felt that the emergence of new general orders, families, and classes required a different kind of explanation. The literature gave rise to a variety of orthogenetic (directed) theories of evolution that usually took refuge in unexplained "autonomous factors" or "vital forces." Most life scientists today reject such theories. The accumulation of paleontological and experimental evidence has strengthened immeasurably the thesis that biological transformations at all levels are explained by mutation and genetic recombination subject to environmental selection. The evidence to support this position is of two types. The first establishes that genetic mutation and recombination of a normal character can produce far-reaching changes. The second deals with the fact that adaptive generalization has generated a more limited paleontological record, thus yielding the impression that novel changes are accidental in character.[3]

THE FAR-REACHING CONSEQUENCES OF NORMAL GENETIC TRANSFORMATIONS

We now know that the idea of macromutations that produce "hopeful monsters" is untenable as an explanation for evolutionary novelties. As expressed by Mayr (1960, p. 355), "Most mutations appear to have only a slight ... effect upon the phenotype. More penetrant mutations are usually disruptive and produce disharmonious phenotypes ... and will therefore be selected against." We need not strain for such explanations. It is now known that many genes can have more than a single effect and can result in multiple (pleiotropic) effects.[4] Furthermore, gene mutations that affect metabolic control functions can have a far-reaching systemic effect. This is especially true of genes affecting the production of hormones by endocrine glands.

It is also now understood that complex correlations link various

[3] The best short treatment of this topic is Mayr (1960). Rensch (1959) and Simpson (1953) are also very good on this issue.

[4] For example, "In domestic chickens, recessive taillessness is a syndrome comprising rudimentation of pygostyle, coccygeal ribs and glands, and tail feathers and deformation of pelvic bones ..." (Rensch, 1959, p. 131.)

aspects of structure to body size and that the growth gradients of body organs are often different from those of the organism as a whole (positive and negative allometry). Thus, a change in body size initiated by a fairly simple mutation may bring with it an enlargement of organs with positive relative growth gradients and the reduction and even atrophy of those with negative relative growth gradients. Sometimes single changes in mass of an organ are sufficient to cross limiting thresholds so that new functional characteristics are realized. For example, the increase in the number of rods in the retina of the eye can reach thresholds that materially alter the nature of pattern and color perception without any change in the basic morphology of the eye. Similarly, an increase in the number of neurons in the nervous system can radically alter the efficiency of internal organic coordination and external environmental response. Rensch (1959, p. 289) points out that "mutations causing positive allometric growth of certain tissues will be especially important, as an increase of cell number may mean the first step of differentiation and the formation of new organs. ... It is especially important to notice that such an increase of tissue may not serve any *special* function at first. Later on ... the superfluous tissue may be 'employed' by a new function in the course of subsequent evolution."

In addition to these considerations, many apparently new structures really are not new. They result from an intensification of established functions and the modification of established structures. "Even when we compare birds or mammals with their strikingly different reptilian ancestors, we are astonished to find how few are the really new structures. Most differences are merely shifts in proportions, fusions, losses, secondary duplications, and similar changes that do not materially affect what the morphologist calls the 'plan' of a particular type ... the evolution of the eye hinged upon one particular property of certain types of protoplasm—photo sensitivity. This is the key to the whole selection process. Once one admits that the possession of such photosensitivity may have selective value, all of the rest follows by necessity." (Mayr, 1960, pp. 208–9.)

Finally, it has also been established that a genetic modification need convey to the organism only a very small margin of advantage for the character to spread through the population in succeeding generations. Fisher (1930) has generated a statistical model of the evolutionary process that shows that a mutation that gives its possessor a one percent advantage over other members of the same species, and which has only one chance in fifty of establishing itself, can sweep through the species in two hundred and fifty generations and might do so in ten.

THE LIMITED PALEONTOLOGICAL RECORD

A second reason why in an earlier phase some biologists found it difficult to accept the evolutionary novelty as a product of the phylogenetic process stemmed from the fact that it seemed accidental in character. It was not adequately explained by paleontological evidence as in the case of adaptive specialization.

Adaptive generalization characteristically occurs when a species occupies a major new adaptive zone in contrast to radiation into subzones or niches characteristic of adaptive specialization. Several features of this phenomenon are of special interest. The environmental change is large scale, and the phylogenetic process is vastly accelerated.[5] In addition, the transition types are restricted in number relative to the populations out of which they issue and the populations that are formed on the basis of their innovation. They are also less frequent in occurrence when compared with the widespread and almost continuous effects of adaptive specialization.

For adaptive generalization to take place, several preconditions must be fulfilled. First, the transition species must have geographical or physical access to the new zone. An aquatic species adapted to a deep ocean niche cannot serve as a phylogenetic bridge to amphibian and terrestrial existence. Second, the transition species must have ecological access to the new zone—i.e., the new zone cannot already be occupied by a partially competing biotype. Third, the transition species must have acquired some minimal preadaptation to the new zone.

To illustrate this last point, we might recall how marine animals that evolved into amphibious and terrestrial types had already evolved an ocean bottom existence in the transition borders between land and

---

[5]Simpson (1953, p. 352) points out that the establishment of new order takes place on the average in much less than a third of the time involved in later spread and diversification.

sea. They had already developed modifications of conventional marine structures to permit crawling. The Devonian period in which the amphibians originated was a time of seasonal drought. When the water dried up the transitional type crawled out of the shrunken pool and scurried overland in the effort to reestablish his aquatic existence. The selective process evolved land limbs in the interest of reaching water, not leaving it. In the process it developed novel organs that ultimately facilitated full terrestrial existence.

Adaptive generalizations, then, are not only less frequent; they involve smaller populations and take place with great rapidity. These three things combined make the paleontological record of the transition types generating evolutionary novelties extremely sparse. It is small wonder that the evolutionary novelties and the process of adaptive generalization were considered by so many to be distinct from the phylogenetic process.

As we have seen, the accumulation of evidence clearly indicates that adaptive generalization is still phylogenesis. We are faced with the fact that the phylogenetic process has generated some highly improbable results, but we now understand that the process itself has specific characteristics that generate improbable results.[6] This understanding does not make the results less awe inspiring, but it does make them somewhat less mysterious, or at least displaces the mystery to a different level yet to be understood.

THE LAW OF THE UNSPECIALIZED

Before leaving this topic it may be useful to underscore one of the large-scale features of the evolutionary process. It is referred to as "the law of the unspecialized" and it asserts that adaptive generalization rarely issues from specialized roots; later species tend to evolve out of relatively unspecialized lines of earlier species.

Specialized lines tend to suffer three disadvantages identified earlier.

[6] Campbell (1965, p. 27) expresses this point well. "Given these conditions, an evolution in the direction of a better fit to the selective system is inevitable. It is through such a process of selective cumulation of the unlikely that the extremely improbable and marvelous combinations found in plants and animals become, in fact highly probable."

(1) They may develop a more restricted range of organism behavior. (2) The genetic pool of the biotype may become reduced in genetic variability. (3) The specializing process may deprive the specialized biotype of relatively undifferentiated or surplus tissues so important in adaptive generalization. All of these factors tend to remove generalizing adaptations beyond the reach of the specialized biotype.

This is an elaboration of one of the preconditions of adaptive specialization enumerated above. It is no accident that minimal preadaptation to a new adaptive zone takes place in the transition zones between two major zones. There may occur here some range of environmental overlap that permits some measure of preadaptation within the old zone and may allow organisms with a wider range of adaptive behavior to exist. In such transition zones the subspecies may retain some interbreeding overlap with other subspecies that supports greater genetic variability. Thus, the unspecialized lines of the parent species are most apt to be found precisely in such a transition zone.

Also worth emphasizing here is another significant aspect of the evolutionary contexts in which adaptive generalizations seem to occur. A necessary precondition for adaptive generalization is the existence of an exploitable surplus in three senses: (1) The organisms will commonly have retained some undifferentiated tissues capable of being employed in the evolution of new structures. (2) The genetic pools will commonly retain a relatively high degree of variability with a wider range of latent characters subject to reinforcement in a modified environment. This is a surplus in the sense that there exists a range of genetic characters not yet pruned by adaptive radiation into narrow environmental niches. (3) There will commonly exist a neighboring adaptive zone where additional resources for self-maintenance are available, if it can be reached.

The stimulus to adaptation and the motivation of the adapting species are the same in both adaptive specialization and adaptive generalization. The stimulus is an incremental change that jeopardizes the biotype by creating a deficit in the resources or some other condition essential to its existence. The adaptation is motivated by the individual and collective striving for self-maintenance and self-reproduction. The adaptations are all in the direction of offsetting the deficit and reestablishing species viability. If the biotype faces a situation where it has relatively differentiated tissues, where genetic variability has been severely pruned by previous specializing adaptations, and where no relief is available in neighboring environmental zones, the only adaptations available to it are fairly limited specific adaptations by what are already highly specialized existing structures and behavioral modes. If these adaptations are not sufficient to eliminate the deficit, extinction will take place.

However, in those relatively rare special cases where environmental change forces a biotype into a deficit position under circumstances in which the preconditions of exploitable surpluses are fulfilled, the drive for self-maintenance and deficit elimination may give rise to adaptive generalization.

This is a way of reformulating the distinction between adaptive generalization and adaptive specialization that will prove useful at several points as we proceed.

The Relationship of Generalizing to Specializing Adaptations

Since both adaptive specialization and adaptive generalization result from the phylogenetic process, it is not surprising to find an important relationship between these two modes of adaptation. Each new general idea opens the way for a new series of ideas. For example, the invention of the wing structure opened the kingdom of the air and an entirely new set of habitats to the class of birds. It supported a whole new set of specialized radiations of form and function.

This illustrates also the relative character of general versus specific adaptations. A new structural type may start out as a generalizing influence relative to previous modes and open the way for a long series of spectacular improvements, only to close in the end upon some range of the environment from which it cannot escape. Thus, the development of the wing deprived the bird of potential hands that could be used to manipulate the environment, and in the interest of flight limited potential size (hence the size of the brain), foreclosing the development of intelligence.

In short, adaptive generalization as well as adaptive specialization can lead to terminating branches on the phyletic tree. The latter is more apt to achieve this effect at the level of species or subspecies, and the former more apt to do so at the level of genera, families, or orders.

57

Biological Evolution Interpreted as a Learning System

The time has come to recall that we began, in the first chapter, with an expression of our interest in the transformation of complex socio-economic systems. It was pointed out that the development or behavioral transformation of such systems (that is, distinct from changes in scale of systems and subsystems) requires that they learn new modes of behavior. We observed that learning can be incorporated in conventional economic models without modifying their deterministic character if the learning is introduced to the system as programmed learning of exogenous origin. Even modified in this way, we concluded, such deterministic models tend to draw our attention away from those aspects of social change that stem from creative, nondeterministic social learning. They fail to explain the sources of development that are endogenous to the system.

The task of understanding the process of development of economic and social systems was characterized as gaining an understanding of the ways in which they operate as creative learning systems. We began this task by summarizing the main features of the modern synthetic theory of biological evolution—having asserted that this process of change behaved essentially as a nondeterministic learning system and might provide us with some initial insights. It now remains for us to indicate more explicitly the sense in which the process of biological evolution illustrates the operation of such a creative learning system.[7]

THE LEARNING SYSTEM AT THE LEVEL OF THE PHYLOGENESIS
OF AN INTERBREEDING POPULATION

We have characterized the process of evolution as a phylogenetic process that takes place at the level of populations of organisms under the combined influence of (1) the collective variability of the characters

[7] Up to this point the author has been able to rely upon a highly developed and well articulated literature. Consequently, the summary presentation just completed is a generally valid and reliable representation of that body of thought, although the organization of the topic may not be exactly duplicated in any other treatment. In the present section, however, where the analysis of biological theory is not based upon a similarly articulated literature, there is a considerably higher chance that the author may have misunderstood or misinterpreted the theory.

in a genetic pool subject to random modification through individual mutation and genetic recombination, and (2) the selective effect of the environment. Under the influence of these two factors the mean characteristics of the genetic pool and their variability (and, hence, the mean characteristics and variability in the behavior and structural forms of the set of organisms) are subject to cumulative modification.

Taken at this level it is obvious that modern synthetic biology describes a process of biological transformation that can be fairly characterized as a creative learning system, or at least a major component of a learning system. Internal to the process as described is a source (mutation and recombination) of new functional ideas representing modifications in behavioral range available to the organism in meeting the challenges of its environment. Under the influence of the process whereby these new genetic ideas are formed and selected for environmental relevance, the behavioral program (genotype) of the organism becomes reprogrammed.

Since these potential changes are random they do not imply or require that the organism itself function as a learning system (although they do not rule it out). For purposes of the phylogenetic process the organism's response is either adequate to the requirements of the environment or it is not. The consequence is a binary on or off—survival or death. Out of the aggregate behavioral variability of the interbreeding population those characteristics that improve the efficiency of organism behavior relative to the environment result in the cumulative modification of the selective behavioral program of a biological population so that the behavior of the organism cohorts is cumulatively modified. Thus, the phylogenetic process describes a system that has the capacity to reprogram itself. The behavior of the population is cumulatively modified through the action of a learning process that is at least partly endogenous and has the capacity to produce behavioral and structural novelties.

At the same time we need to recognize that this representation of biological evolution as a learning process is incomplete. For one thing, we still have an insufficient idea of the process that generates new genetic ideas to offer a satisfactory explanation of the internal source of novel behavior at this subsystem level.

More important for our purposes, the description of the phylogenetic process at the level of the interbreeding population considers the influence of the environment to be an exogenous factor. The environment acts as a selective screen to sort out biological experiments into those that are relevant and those that are not. It is recognized that, in the absence of changes in the warp and woof of that screen, all biological changes would eventually come to a halt, even though the process of mutation and genetic recombination continued full force. In consequence, a great deal is made of the importance of changes in the environment. However, *such changes are not characteristically represented as a function of the phylogenetic process.* The sources of environmental change are considered to be exogenous to the process. The phylogenetic process thus described is an incomplete representation of an endogenous learning system.

This assertion requires some elaboration because the concept of the environment is so fuzzy that the meaning and significance of this statement may not be immediately apparent. It will help us here to make a distinction (following Bates, 1960) between the *operational environment* (the sum of the phenomena that directly impinges upon the organism and population) and the *potential environment* (the sum of the phenomena that might conceivably impinge upon the organism— a sort of "total reality" frame).

In a limited sense the description of the phylogenetic process does make the operational environment a function of the process.[8] When discussing earlier the two modes of adaptation, we observed that the process of adaptive specialization may reduce the proportion of the potential environment that forms the operational environment for the population. In short, the equilibrating, stabilizing series of adaptive specializations not only tends to improve the correlation of organism behaviors with the operational environment, it also tends to reduce the adaptive range in a way that may narrow the operational environment. Similarly, we observe that adaptive generalization has a tendency to expand the operational environment because of the improvements in the adaptability of the organism.

However, this representation of the operational environment and the

[8] Although this functional link often remains implicit in the conventional treatments in the literature.

population of organisms as intersecting and interacting systems provides no explanation of the changes in potential environment that are essential to explain the continuing operation of the phylogenetic process. It does not deal with the possibility that the behavioral transformation that characterizes an adapting population may have some effect, in turn, upon the potential environment.

That such direct and indirect interactions between adapting populations and changing environments do occur cannot be doubted. As biological evolution has proceeded some populations have come to form part of the operating environment of other populations. The interdependence of biological species has long been known to biologists and studied by them. The forms of interdependence can be classified into three general categories (Bates, 1960). (1) There are food relationships. These represent a wide variety of ways in which species depend upon other plants and animals for food. They can develop into extended food chains. (2) There are structural relationships. Some species depend upon other plants and animals for physical support, or shelter from enemies, or protection from climate, etc. (3) There are reproductive and dispersing relationships. The relationship between insects and flowering plants is an example.

The discipline of bioecology concerns itself with these phenomena. Is it not plausible to think that this discipline might provide the basis for a more general characterization of the process of biological transformation—one that would link adapting populations and changing potential environments into an internal feedback relationship characteristic of a more complete endogenous learning system? We turn to examine this possibility.

THE LEARNING SYSTEM AT THE LEVEL OF THE ECOSYSTEM

Ecologists do move directly to a consideration of operational environments internal to the adaptive system. They characterize the system as an ecosystem and concern themselves directly with what they describe as community relationships. The community is conceived as a set of functionally interrelated biological populations such that the operational environment for each differentiated population is defined in part by its functional and structural ties with the other populations.

61

Their adaptations in form and function or fluctuations in number can lead to a transformation of the operational environment of each species. One might suppose that the study of these coadaptive feedback relationships will lead us to a more general representation of the evolutionary process.

However, bioecology carries us only a limited way along this path. The principal reasons stem from its preoccupation with the coadaptations of functionally related populations. Since the externally linked populations form an important part of a given population's operational environment, the ecosystem defines a system of coadaptive operational environments. This carries us a step beyond the characterization of the relationship between adaptation and the operational environment at the level of the interbreeding population. There we were restricted to observing that population phylogenesis could narrow or broaden the range of the immediate operational environment. In ecological coadaptation the operational environments are, themselves, subject to changes in character as well as changes in range or scale. All of this is endogenous to the ecosystem.

This preoccupation with the coadaptation of operational environments has resulted in the fact that phylogenesis operating at the level of the ecosystem is viewed almost exclusively as a process of mutual adaptive specialization. The interdependent biological community of the ecologist is always characterized as a stable equilibrating system, even though operating at a higher level of system abstraction and heterogenous aggregation. It does not deal with the disequilibrating adaptations we know as adaptive generalization.

The reason for this can be derived from our understanding that adaptive generalization occurs when certain special conditions are fulfilled which allow the population to move into a new adaptive zone. Reinterpreted, we might say that adaptive generalization has, among other attributes, the characteristic of transforming a potential environment into an operational environment. The operation of the ecosystem, as traditionally conceived, does not deal with the relationship between phylogenesis and changes in the potential environment.

The result is that bioecology has been absorbed almost completely with descriptions of the functional organization of the ecosystem. It has devoted its energy to the task of defining the functional interrelationships that characterize such ecosystems as the northern hardwood forest, the tropical rain forest, the savannah grasslands, etc. As Bates (1960, p. 548) points out, "The ecologists have been very busy with the enormous task of analyzing things-as-they-are, so that they have little time or energy left to worry about how they got that way." In short, ecology has not analyzed the dynamic interaction between populations and their environments under the influence of phylogenesis that led, through geological time, to the formation of the different types of ecosystems.

The ecologists do describe a dynamic process that may be mistaken for this by the casual student. It is the sequence of biological species that may follow one another during the process of reestablishing an ecosystem where it may have been destroyed in a particular locale. Here, a distinctive interdependent sequence is followed until "ecological dominance" is reestablished. It would be erroneous to infer from this, however, that the sequence is the same as, or even similar to, the phylogenetic sequences that led to the formation of the ecosystem. The restoration of dominance described by the ecologist explains the way in which a stable set of ecological interrelationships is reconstructed out of biological materials representing already developed species. It does not involve the phylogenetic process as a constituent part of the mechanism.[9]

Bates (1960, p. 566) expresses it this way: "Ecologists have devoted a great deal of study to community succession. The relevance of this study for community evolution is somewhat indirect. . . . Succession, I think, is analogous with wound-healing in an organism, not with evolution." He goes on to point out that the modern synthesis in biology is incomplete and requires the more complete development of what he calls paleoecology. The evolutionary history of communities or ecosystems remains to be properly articulated.

---

[9] Boulding (1966a) is one of those who has seized upon the principle of biological succession or the reestablishment of ecological dominance as a possible model of social change. However, it should be clear that it is not a creative learning model in any sense. Indeed, it is closely analogous to the concept model that visualizes programmed learning as a deterministic stage sequence, as was discussed in the first chapter. It is not clear that those who have been attracted by it understand its limitations in this regard.

We conclude, therefore, that phylogenesis operating at the level of the ecosystem does not form a model of the learning system that makes changes in potential environments endogenous to the system. As a consequence, it cannot deal successfully with adaptive generalization or the disequilibrating mode of a learning system. Furthermore, it does not carry us beyond the traditional descriptions of the phylogenetic process that see adaptive generalization as an opening, disequilibrating, displacing, generalizing step. It fails to explain the conditions under which one generalizing adaptation might set the stage for another in a cumulative series of adaptive generalizations. Indeed, the literature on phylogenesis places greatest emphasis on how adaptive generalization turns into adaptive radiation. Behavioral displacement is viewed as transmuting in the next round into behavioral specialization.[10]

THE LEARNING SYSTEM AT THE LEVEL OF THE BIOSYSTEM

The biological literature does recognize quite early the cumulative effect of adaptive generalization. It can be observed that, over the span of evolutionary history, a kind of progressive cumulation of improvements in organism adaptability took place. However, this generalization rests upon the observation of evolutionary trends at the level of the biosystem as a whole. The trends are said to be progressive—hence the origin of the term "progressive evolution."[11]

Thus, the progressive trend in adaptability is a manifestation of the operation of the phylogenetic process at the level of the biosystem, rather than at the level of the interbreeding population or ecosystem. Adaptive generalization is seen as a rare occurrence in any one phylogenetic line. To see the progressive trend in the series of adaptive generalizations, one must back off far enough to see the results of the multiple lines of phylogenesis in the biosystem taken as a whole. Thus, progressive evolution as a trend can be observed by noting the progressive advances in adaptability displayed, not by a single lineage but

[10] This helps to explain why human ecology, which grew out of the images associated with bioecology, has made very little contribution to understanding the process of social change.

[11] As we observed earlier, these progressive trends were initially held by many to be unexplainable in terms of the process of phylogenesis. Today the burden of the modern consensus is that phylogenesis explains both improvements in behavioral adaptation and behavioral adaptability.

by a succession of dominant species in successive evolutionary eras. It is pointed out by the biologists that these adaptive generalizations can be cumulated in an open-ended progression that contrasts with specializing adaptations because the improvements in adaptability, once developed, tended to be maintained by all subsequent animal phyla. (See Rensch, 1959, p. 285.)

It is true that the long cumulative trend of adaptability that draws the attention of the biologist does not characterize each of the multitude of phylogenetic lineages in evolutionary history, and is an accurate characterization only at the biosystem level. At the same time, it is necessary to recognize that, in the single case of human evolution, the phylogenetic lineage represents a nonterminating series of improvements in behavioral adaptability. There are other instances in evolutionary history when a single line succeeded in stringing together several adaptive generalizations before running up against some limiting factor.[12]

It is important to emphasize that the description of these progressive trends does not offer an explanation of the way the system operated to sustain cumulative advances in organism adaptability. To say that the trend is observable, and that it rests in some sense upon the fact that the selective advantage of improvements in behavioral adaptability tends to preserve these characteristics in genetic pools, does not explain why these improvements continued to cumulate. The analysis of phylogenesis tells us how the adaptive generalizations can arise out of phylogenesis when the biotype invades a new adaptive zone. But it offers no explanation of how the new adaptive zones arise. The literature does not characteristically visualize the environmental changes as a product of the evolutionary process itself or inquire specifically into the kinds of environmental changes that are more apt to generate improvements in adaptability rather than improvements in adaptation.

Expressed another way, the literature has been absorbed with the "fitness" of the population. That is, it has concerned itself with the way the genetic pool and associated organism behavior fit themselves to the environment. It has not characteristically raised questions about the "fitness of the environment." It is of central importance in the explanation of progressive evolution to inquire into the role of the environment—its "fitness" to support adaptive generalization and how

[12] See p. 55.

such "fit" environments may be generated by the evolutionary process. Barring some understanding of how the process works with environmental change considered internal to the system, we cannot successfully specify the character of the learning system at the level of the biosystem. Explanation stops short with the limited learning system of phylogenesis.[13]

At the same time, we can see the major features of a more general process sufficiently clearly to feel confident that progressive evolution at the level of the biosystem does, indeed, contain endogenous processes that can generate new adaptive zones and sustain adaptive generalization.

A systematic change has been taking place in the environments shaping the selective process in phylogenesis that has, itself, been a product of the evolution of the biosystem. Indeed, as that process has proceeded, the relevant concept of the natural environment (from the point of view of the biological organism or the adapting population) has itself become transformed. Later adaptive phases were shaped by genetic response to an environment that consisted, in considerable measure, of attributes that were themselves a product of biological evolution. Biological organisms come to form an ecological environment conditioning further adaptation.[14] What is one species' heredity becomes another species' environment. Subsequent adaptations are conditioned by the existence of prior adaptations. Thus, a larger and larger portion of the modifying environment consists of other evolved (and evolving) life systems and the relative transforming role of the changes in the cosmological-geological-atmospheric environment becomes progressively diminished. The vital energy and material resources of the earliest species, and the features of structural support, were obtained directly from this lifeless environment. These sources for later species became the intervening forms of earlier species.

---

[13] A fairly careful search of the literature gleaned a single citation in this area— a book by Henderson, *The Fitness of the Environment* (1927). However, this book deals exclusively with the fitness of the total cosmological environment and the fitness of its physical chemistry to support life systems and does not come to grips with the issues being raised here.

[14] We have already noted the nutritional, structural, and reproductive interdependencies that form the network of interlinking and interacting environments that characterize the ecosystem.

In short, the environments that shape the selective process become progressively redefined in the process of evolution itself and the role of inorganic (nonreproducible) components becomes progressively diminished relative to the organic and biological (reproducible) components. In the later phases of evolutionary history the major portions of environmental change forming the selective process are internal to the process of evolution. It can fairly be characterized as a largely endogenous learning system at the biosystem level.

It may also be a valid generalization to say that the environmental change that has been generated by the operation of biological evolution has tended progressively to generate new broad adaptive zones of the type that favor adaptive generalization.

The expansion of the biosphere, with its splitting into myriad differentiated lineages and their reticulation into functional ecosystems, has the effect of multiplying and continuously modifying the zones and niches occupied by the differentiated species. In the process new zones are created that require adaptive generalization on the part of some species for their exploitation. For example, the earliest land animals could not evolve until the land had been previously invaded by plant life capable of supporting animal life. Once this was accomplished, a broad new adaptive zone existed that possessed the potential for occupation by animal forms. Such broad new zones as these tend to require adaptive generalization for their exploitation. At the same time, each adaptive zone generates species on its fringes that are preadapted to the new environmental zones that develop as near neighbors, geographically and behaviorally. We have seen that these transition types are most apt to have the behavioral and genetic characteristics that will favor adaptive generalization. At the level of the biosystem we can see that the process of evolution, in addition to generating new species, generates new adaptive zones and creates conditions that favor bringing some species to transition positions where the new zone is within reach of their phylogenetic potential.[15]

---

[15] At this writing there is only one other source where the author has found something like this point. Campbell (1965, pp. 39–40) makes a similar point almost as an aside. "Over the whole course of evolution, the simpler combinations come first, and it follows that the great bulk of adjacent ecological niches are in the direction of greater complexity. ... the mean complexity of all occupied niches steadily increases from epoch to epoch."

Thus the evolution of the biosystem is a learning system exhibiting the property of transforming and reorganizing its own behavior. Internal to the operation of the process itself are process characteristics that create conditions that occasionally encourage phylogenesis to stretch the adaptability of the population. The operation of the feedbacks at this level is not well understood because biological scholarship has not dealt altogether successfully with this problem.[16]

[16] Even where someone, like Bates (1960), begins to call for an extension of biological scholarship into this realm, it is not clear whether his concept of the evolutionary history of the ecosystem envisages only an attempt to determine how certain identified equilibrating ecosystems came to be developed through the operation of the phylogenetic process, or whether the concept is broad enough to include a study of the way interspecies relationships may have played a role in sustaining the cumulative effect of adaptive generalizations.

In this connection it is worth mentioning that a considerable search of the literature on bioecology yielded a single additional source that even treated the subject of ecology and evolution. Allee et al. (1949) includes a large section chiefly devoted to outlining the way certain kinds of microenvironments contribute to genetic isolation and natural selection. This treatment is not different in basic character from that given to the same material by such evolutionists as Rensch.

However, in a short section devoted to the "Evolution of Interspecies Integration and the Ecosystem" the process is identified (p. 695) as "the genetic modification of the ecologically associated organisms in relation to each other, in the aggregate, resulting in the evolution of the community as a whole." This is distinguished from the traditional concepts of "succession" and is characterized as the "phylogeny of the definitive grouping of species within the community."

Here the basic distinction is nicely drawn and this phylogeny of the community is treated in some detail and summarized as follows (pp. 728–29): "... reciprocal genetic patterns evolve by means of such selection and produce interspecies adaptations, interdependence and integration. Harmful disoperation between species eliminates itself. Exploitation tends to evolve toward toleration of mutualism. ... The evolution of the division of labor and integration between species results in a biotic system that may appropriately be called an intra-species supra organism. The incorporation and control of the physical habitat by the intra-species supra organism produces a unitary ecosystem. Homeostatic equilibrium within the ecosystem is in large part the result of evolution."

This last sentence reveals the bias alluded to above. Ecologists tend frequently to equate the ecosystem with the biosystem. And in the rare instances when they view ecology in the context of the "phylogeny of the community" they reveal themselves as absorbed with the interacting adaptive specializations that yield a stable classical ecosystem. It seems that they do not even raise the question of the environmental feedbacks that foster adaptive generalizations.

68

BIOLOGICAL EVOLUTION AS A STOCHASTIC LEARNING PROCESS

When we see the effects of the operation of the phylogenetic process at all levels from the interbreeding population to the total biosphere, we can see that most of the factors leading to behavioral transformation are operating endogenous to the system. There is no evidence to support the notion that exogenous or orthogenetic forces are at work to establish a predetermined result for the process at any level. This is clearly not a manifestation of programmed learning. It is a creative learning process.

We know that the process is nondeterministic in character because knowledge of the phylogenetic evolutionary process does not support positive prediction of the future course of evolution. (See Goudge, 1961, pp. 121–27.) Knowledge of the process does not allow us to say that a given phylogenetic lineage will develop in a certain way in the future.

The capacity of the biosystem to create new modes of behavior in life systems is a stochastic process. The mutations and genetic recombinations that generate new behavioral ideas at the level of the interbreeding population are constrained. They are conditioned by their genetic lineage. At the same time, the history of the biotype leaves open a number of novel behavioral options. We have observed that the range is larger when greater variability exists in the genetic pool. Which of these options becomes reinforced through selection depends upon the character of future environments encountered by the biotype. The nature of these future environments is, likewise, constrained. It is conditioned by the history of the cosmos, atmosphere, geosphere, and biosphere—with the latter assuming the dominant influence in recent biological history. At the same time, that history leaves open the possibility of a variety of future environmental states.

The emergent novel behavior is the product of the interaction of two sets of conditional probabilities—one related to emerging behavioral options and the other related to potential future environments. These two probability sets are interrelated. Each event, in the chain or sequence of events that marks historical development, modifies both sets of conditional probabilities and establishes a new probability matrix within which the next event in sequence will emerge. It is rather like a complex, open-ended Markov-chain process.

69

A similar metaphor for this process is offered by Slobodkin (1968) when he likens evolution to an "existential game." It is a game which the population plays with the environment under the rule that no player can leave the field of play and the "bizarre meta-rule" that any player that quits the game is killed. The measure of a player's success is his persistence in the game. The optimal strategy for any player is to maximize his homeostatic ability. "If organisms are, in fact, maximizing their homeostatic ability during the course of evolution, it must be the case that they respond to environmental perturbations in such a way that they not only minimize the departure from steady state conditions caused by the perturbation but also maximize their ability to withstand further perturbations." In short, selection favors increases in adaptability as previously discussed.

One might suppose that in this model the positive prediction of future states might be replaced with at least a probability prediction. But this model does not carry one very far in making even probability predictions about the future course of biological evolution. The kind of empirical information necessary to give dimension to these conditional probabilities sufficient to attach probabilities to specific future states appears to be beyond the possibility of acquiring.[17]

We have now achieved one of the objectives we set out for ourselves in chapter I. Through the study of the modern synthetic theory of biological evolution we have arrived at the specification of a concept model of a process of behavioral change that can be characterized as creative learning. We have learned that it is a stochastic process or game process that does not support positive prediction. This raises two important questions: Is it possible that such a concept model may

[17] This is not to say that a kind of conditional probability prediction directed to predicting process characteristics (such as the likelihood of extinction or a high rate of adaptation, given certain population-environmental characteristics) or that a kind of negative prediction (that renders evolutionary theory empirically testable) is not possible (Goudge 1961, pp. 70–78, 121–27). But these are methodological matters which will be given greater attention in chap. IV. There is some possibility for making more useful probability predictions in more limited biological domains. Classical genetics at the level of microbiology has long made possible useful probability predictions about the effects of genetic recombination brought about through selective breeding, etc.

prove useful in social analysis? In what way might such a model yield a more general approach to prediction and planning?

Similar models are not in widespread use in social analysis, but they are not altogether missing.[18] The game theory model is an explicit instance of a stochastic decision model as are the models of modern statistical decision theory. These models can be viewed as frameworks in guiding planners in the placing of bets based upon a set of conditional probabilities. It is safe to say that they have found successful application only in limited circumstances.

There may be something like this concept implicit in the growing interest in technological forecasting among development planners. To date this term seems to embrace a variety of meanings. By technological forecasting some seem to have little more in mind than a kind of exogenous system search to identify little known techniques that can be used to reprogram a machine system open to information. However, the dominant theme seems to be a desire to identify potential emergent lines of new technology (and, hence, new environments and potential new endogenous behavioral modes). Such an anticipatory search can be viewed as a process of attempting to anticipate some of the most probable outcomes from a set of conditional probabilities. Much attention is devoted to the history of technology, the most advanced current work, and to the most pressing technological problems in the effort to develop some idea of the empirical dimension of these conditional probabilities.

As has already been implied in the discussion of biological evolution, the application of the concept of the stochastic learning model implies

[18] Among the most highly developed concept models of this type are those exhibited by the recent development of mathematical learning theory within the tradition of behavioristic psychology. (See Bush and Estes, 1959.) As expressed by Estes (p. 11), these theories define learning "... in terms of trial-to-trial changes in the conditional probabilities of responses to stimuli, the laws of learning take the form of rules specifying the probabilities of these changes under various experimental conditions." Learning theory is presented as a stochastic process of Markov-chain type whereby the stimulus environment and the response probabilities of the learning organism are interrelated in identifying the possible emergent sequences of events. An example of an attempt to apply the stochastic model to economic analysis can be seen by examining Murphy (1965). It is not appropriate to our purpose to explore these models in greater detail at this point.

a different approach to prediction and planning. Prediction is not positive. At best it is probabilistic and conditional. Planning in such a case cannot seek to preadapt the system to a firmly anticipated future state. The planner is faced with a range of possible future states (problems or opportunities), some of which seem more probable than others. Under such circumstances planning is more absorbed with organizing and managing system behavior in a way calculated to minimize departures from a steady state and to keep the system flexible in meeting future problems or exploiting future opportunities. Excessive specialization or closure on existing modes might easily constitute an evolutionary trap of the kind so familiar in biological evolution.

Prediction and planning under the control of this model concept is, indeed, more general than that characteristic of the concept of the machine system open to information. This sketch should make readily apparent, however, that such a concept still does not adequately characterize the developmental process that we observe in social systems. One might speculate that it could be brought closer to the social process by enlarging the concept of planning. Planning might be viewed as directed to changing the probabilities of the future events or future emergent modes of behavior. But such a change in orientation immediately carries us beyond the characteristics of the stochastic learning system revealed by the process of biological evolution. That simple step transmutes the model concept into a still more general creative learning system. It may still retain some stochastic properties but display different process characteristics. Hidden in such a seemingly minor modification in the planning objectives is the presupposition that the planner could have both the ability to modify the probabilities of future events and a reason for doing so. These capacities are unique to the social process and cannot be explained or carried out within the framework of a stochastic learning model revealed by the study of biological evolution.

There is a more general creative learning model. In this volume it will be characterized as a process of *social learning*. A fundamental task for social science is to articulate and apply this model. The biological analogue will not serve. However, it provides us with a model that illuminates those social processes that are usefully analogous to the biological model, and it helps to highlight those characteristics that are unique to social learning. Its use for this purpose is undertaken in the next chapter.

Summary

The process of biological evolution is based upon a complex function of mutation and genetic recombination under the influence of environmental selection. This phylogenetic process displays two modes —adaptive specialization and adaptive generalization. The former acts to improve the adaptation of stereotyped organism behavior. The latter acts to improve the adaptability of the organism. The one may narrow the operational environment of the species while the other tends to broaden it. Both modes of adaptation are manifestations of phylogenesis, and adaptive generalization usually opens the way for a whole new round of adaptive specializations.

The process of phylogenesis displays the characteristics of a learning system. It describes the process through which the behavior of a biotype becomes transformed by virtue of the biological population's internal capacity to generate new behavioral ideas through interaction with the environment. This learning process exhibits both the capacity for behavioral refinement and behavioral innovations.

The innovative transformations are of special interest to us because they are manifestations of a learning system performing as a developmental system. We note that the adaptive generalization acts to increase the complexity and improve the rational organization of organisms and this is the morphological counterpart to the development of more adaptable behavior. Adaptability is fostered by a series of improvements in environmental tolerance, organism mobility, and discriminatory capacities of the organism. The latter, as manifest in the development of the nervous system and the brain, has been especially important in the evolutionary history of adaptable behavior.

We observe that the phylogenetic process, as conventionally described, falls short of internalizing all of the creative elements of self-reorganization or behavioral reprogramming. However, we can see that, at the level of the biosystem as a whole, many of the environmental changes that form the shaping edge of the creative dialogue are themselves a product of biological evolution. We can visualize them as

73

components in an internal feedback response that makes of the evolutionary process an endogenous learning system.

This aspect of the process cannot be satisfactorily detailed because it has not been adequately attended to by the life scientists. However, the nature of the process is sufficiently well known to characterize it as a stochastic process that does not support positive prediction. It turns out, therefore, that it does not provide a fully adequate model for describing social change, nor does it form an adequate base for the conduct of social prediction and planning. Nevertheless, it may illustrate some aspects of this process and, as a base for comparison, serve to highlight some of the unique characteristics of the social process.

III   Social Evolution

WE HAVE COME TO THE POINT that we recognize that economic and social development implies changes in modes of behavior. This, in turn, implies that development is essentially a learning process. Learning processes can be conceived as taking the form of programmed learning or creative learning. In the interest of investigating model concepts that can be characterized as creative learning, we have investigated the stochastic learning model represented by the modern synthetic theory of biological evolution. That review makes plain that there is another creative learning model that we characterize as social learning. In order to give this model a more complete articulation, let us examine the principal features of the process of social learning and the way in which it has emerged.

Consider, first, the way in which the process of social learning came into being. This process has both an individual aspect and a group aspect.

The Dawn of Social Learning

THE LEARNING SYSTEM AT THE LEVEL OF THE ORGANISM

The most striking thing about evolutionary history is the fact that the operation of phylogenesis in its generalizing mode has created improvements in organism adaptability until it has generated learning organisms. This gives special point to one of the large-scale features of evolution already emphasized. We observed earlier that instead of modifying the genetic base of behavior by pruning and reshaping a stereotyped pattern, adaptive generalization, by promoting adaptability, provides the organism with the power to *modify its own behavior* during its life cycle *without sole recourse to another round of selective transformation of the genotype.* We observed that this power was particularly served by the development of a nervous system and brain that permits discriminating behavioral responses, and that this power has reached its highest manifestation in man. These traits persisted and developed because, once invented, they enormously enhanced the survival characteristics of the organisms and populations so favored and became inscribed in the genetic base of the species.

Expressed another way, the behavioral reprogramming of organism behavior that is phylogenesis gradually evolved a program (genotype) that provides the organism with the power to reprogram itself—to act as a true learning system at the organism level. Phylogenesis operating as a learning system produced a learning subsystem—the learning organism—that operates at a different level and by more direct means.

The development of this new biological capacity did not create a way for the new adaptive behavior (acquired by the organism in the course of its lifelong encounter with the environment) to be passed on to its progeny through the genetic material. Thus, during the earlier phases of the evolution of this discriminating process, there was no means for accumulating acquired behavior and each organism had to "rediscover" the world for itself.[1]

As these learning powers became enhanced in later species, we could see the emergence of a new learning dynamic. As organisms acquired the power to perceive their environments, interpret those perceptions, and generate a feedback response, they found that their environments included other members of their own species engaged in a learning response to the environment. In time, the power of mental abstraction arrived at the point where the behavioral responses of others could be perceived and interpreted in a way that permitted behavioral mimicry. This opened the door to the *accumulation of acquired behavior* in

---

[1] The neo-Lamarckian theory of evolution maintains that these acquired characteristics come to be reflected in the genetic code and are thus transmitted in standard genetic fashion. This theory is widely discredited in the life sciences today. The most persuasive evidence points the other way.

a population not subject to the mortality constraints of organisms. In short, the learning capacity of the organism became socialized into a more general learning system that operates once again at the level of the population rather than the individual. But the process at work here obviously exhibits a different dynamic form than that of phylogenesis.

The advantages of shared learning for behavioral adaptation and survival assured the reinforcement and development of this mode of learning.[2] It reached its peak in man, in whom the power of abstraction is raised to a level supporting formal symbolic modes of communication or sharing of acquired experience. In the human species the learning organism reaches the point where learning becomes largely socialized because the dominant aspect of the individual organism's learning environment is the presence of and the sharing with other human learning organisms.

The development of socialized learning opens the way for an important change in the way learning systems operate. The phylogenetic process always operated through genetic differentiation. Under the influence of variations in environmental ranges genetic pools became progressively differentiated into subsets (the process of speciation). In a few instances the rudiments of genetic diffusion were present. This is a process that brings about the transformation of biological systems by the diffusion of genes between species through introgression or hybridization. Where this occurs it leads to a convergence of species characteristics rather than the divergent characteristics of adaptive radiation. This mechanism has played a greater role in the evolution of plants than animals, but its role for the most part has been extremely limited in both. It is obvious, however, that the process of socialized learning places great reliance upon the process of information diffusion with its attendant convergent qualities.[3]

---

[2] The evidence indicates that elementary forms of shared experience and social behavior appeared in the higher animals even before the evolution of man. The so-called social behavior of some insects like bees and ants is not based upon the interaction of adaptable organisms in a true social process but is a complementary behavior pattern frozen into the genotype of the species.

[3] And, interestingly enough, as this facility for social learning evolved in mankind, it had the effect of reinforcing the effect of genetic diffusion in this evolutionary

THE LEARNING SYSTEM AT THE LEVEL OF THE SOCIAL SYSTEM

The socialized learning of the individual human organism gives rise, in turn, to another principal feature of social learning. Socialized individual learning goes beyond merely sharing a range of acquired individual behaviors through communication. This sharing of experience has created the opportunity for individuals to amplify the components of individual behavior through cooperative group action. Living in a social environment tends to foster the sharing of activity as well as information.[4] This has led to social behavior at a higher order level of system complexity—the social system.

Some aspects of group behavior find their origin, of course, in the necessities associated with biological reproduction and the nurture of the young. But group behavior that goes beyond the level of the family as a social group is most often based upon cooperative behavior essential to the amplification of the components of individual behavior—an extension of biological morphology and physiology.

At first the behavioral amplification took the form of pooling or aggregating common characteristics of behavior. Man, faced with a task that exceeded individual strength (e.g., moving the carcass of a large animal), learned to couple multiple units of manpower through social action. Just as man learned to amplify his biological effectors through joint action, he learned to amplify his biological receptors. Through shared perception (e.g., the use of scouts in a hunting or collecting party) he acquired the power to inspect or encompass a broader range of experience. He also learned to amplify his biological

---

line. Organism adaptability was raised to such a level in *homo sapiens* that the adaptive zone of the human species developed no subzones or niches sufficiently restricted to permit the development of completely isolated subsets of the genetic pool. This species has remained one large interbreeding population so that a large behavioral range is superimposed upon a diverse genetic pool that fosters genetic diffusion through migration and interbreeding. The extreme (by historical standards) behavioral adaptability of the organism supplemented by shared experience through social learning ameliorates the influence of the environmental characteristics that tend toward population differentiation.

[4] Indeed, modern psychology and epistomology link the idea and the act. The symbolism of language and symbolic ideas are born in action. Thinking is seen as imagined action. See George Herbert Mead (1934, 1932).

interpreters through joint action. During the early phases of social life the pooling of memory appears to have been important. The possibility for these forms of shared action is based upon and enhanced by the fact that the phylogenetically evolved characteristics of the human species are associated with a biological predisposition to social or group behavior. The forms of social organization associated with these early forms of pooled behavior were relatively simple.

As social learning advanced, man learned to develop human artifacts that further amplified the components of behavior. He developed tools that extended his powers (as the club was an extension of the arm). He learned to exploit external energy sources in the form of the work animal, the kinetic energy of water and wind, and, ultimately, combustible energy. He developed the artifact of written language and number systems that extended the power of memory and the efficiency of conscious thought. But whatever the form of behavioral innovation, it typically required social action for its exploitation. Such social action extended beyond the aggregation of homogeneous human characteristics to the functional integration of specialized activities. This created higher order and more complex social groupings.

Just as the individual human organism has the phylogenetically evolved capacity to modify its own behavior through acquired experience, the social system or group has the capacity to modify group behavior. Both the individual and the group embody the capacity for creative learning. Learning and behavioral transformations take place at both levels and both are essential aspects of what is referred to in this book as social learning. This embraces both the socialized learning of the individual and the individually supported learning of the social system. Thus we see that the learning process that is biological evolution has spawned still another learning system. By means of a nonphylogenetic process for accumulating information and transforming behavior, it has produced social evolution.

It is this process of social learning in which we are behaviorally immersed and which generates the phenomenon of social development that forms our primary concern here. We now need to determine the ways in which our understanding of social learning is informed by our knowledge of the biological process antecedent to it. We need to determine the characteristics of social evolution that are unique to it and

79

find no counterpart in the phylogenetic process. We need to understand how these processes work to generate social change and how these changes are related to the problems of social organization and social behavior that occupy so much of our attention. The task is obviously larger than this book and longer than this generation. But enough is known today to permit us to sketch the gross features of this domain.

A Comparison of Social Evolution with Biological Evolution

In this section we turn to the analysis of social evolution against a conceptual backdrop provided by the analysis of biological evolution presented in chapter II. Our aim is to determine the ways that the social process is similar to or different from the biological process, a task which we shall first approach from the perspective of the socialized learning of the human organism.

SELF-REPRODUCTION AND SELF-MAINTENANCE

In social evolution, as in the biological phase, the necessary conditions are self-reproduction and self-maintenance. Basic to both processes is the replication and transference of coded information governing the behavior of organisms and behavioral systems.

In the case of biological evolution the information is embodied in genetic codes written on DNA and embedded in the gene structure of the organism. It specifies the form and behavioral mode of a mature member of the species and of the stable ecosystem of which it is a part.

The nature of information in the social phase forms the first contrast. It consists of information about man's acquired experience—information concerning adaptive behavior "learned" by organisms during their life cycles. This includes the acquired forms of cooperative group behavior. Self-reproduction in this realm consists in the replication and transference of a body of acquired knowledge—the "handing down" of experience.

The replication and transference of acquired experience require that it, likewise, become embodied in some material form. With the emergence of the human species, the generalizing process of cerebralization crossed a threshold of capacity that permitted the codification of acquired information. This did not take the form of chemical coding in biological genetic material, but took the form of language—gestural at first, then oral, then written. With man came the cerebral capacity for abstracting essential to the development of symbols. Numerous material media for their embodiment (such as artifact writing) became available. When the symbols and their idea freight could be passed on to successive organisms, the acquisition of experience became cumulative. The human species as a population came to acquire a new kind of information pool similar to the genetic pool in the biological phase, yet different. The differences, however, are sufficient to give rise to a different process of change even though some generic characteristics remain.

THE PROCESS OF SOCIAL TRANSFORMATION:
SOCIOGENESIS OR SOCIAL LEARNING

Each evolving human population and each of its identifiable subpopulations possess a more or less integrated pool of acquired ideas and an associated set of behaviors making these concepts manifest in action. This is the essence of culture. This pool of ideas is maintained and reproduced through the operation of the social process. But, as in the biological phase, the idea pool is not only reproduced, it is transformed. In the process the content of culture and the organization and behavior of social systems become transformed.

There are other similarities. In phylogenesis, genetic innovation is the result of a process of genetic mutation and recombination. In "sociogenesis" we witness the persistent force of idea invention and recombination. These socially acquired ideas, like genetic mutations and recombinations, contain the potential for changes in behavior. Like biological mutation, human "idea mutation" does not always generate relevant ideas. Those idea inventions or behavioral innovations that are not consistent with the interplay between operating environment and operating goals tend to lose force. Those that promote a convergence between environment and social goals are reinforced.

81

Thus the variety of novel ideas and behaviors is sifted to generate a residual information pool and a tested set of behaviors in a way not altogether different from the phylogenetic process. Sociogenesis, like phylogenesis, is engaged in maintaining, reproducing, and transforming information and behavior that is relevant to environment and goals. A process heritage that carries over from phylogenesis into sociogenesis is clearly shown. At the same time, the very act of identifying these similarities calls our attention to striking differences revealed by the process of social learning.

The *first* and perhaps most fundamental difference has been suggested early in this chapter. The invention of new ideas and the innovation of novel behavior in the social realm are not a manifestation of stochastic changes in the molecular biology of the germ plasm of the organism. They are the product of the active, conscious, discriminating capacity of a learning organism. They are the product of a cognitive process that perceives environmental objects and events and interprets the relationship of these perceptions through exercise of a generalizing, associative, analytical faculty. This basic difference in the origin of the creative idea is the basis for a set of additional distinctions between phylogenesis and sociogenesis.

The *second* has also been mentioned. The perceiving, discriminating organism encounters in its operating environment other organisms capable of making learning adaptations and with whom it is able to communicate and share ideas and behavioral modes—including those ideas and behavior modes that form the basis for cooperative group behavior. Thus, individual learning becomes socialized. The transfer of information is not restricted to a finite moment when the gametes of two organisms of opposite sex form a union as in the biological process. Knowledge becomes transferable outside the rigid, wasteful, discrete time lags of the genetic mechanism. The acquired knowledge of human organisms can be passed on, not just to progeny but to others in the same generation. The possible lines of information transfer become permuted into a virtually infinite set. In this new setting the diffusion of information is favored. A strong tendency develops for the merging and interaction of the subsets of the information pool. This forms a marked contrast to the divergence and isolation of the elements of the gene pool common to biological evolution. The feedback between operating environment and organism behavior is much more direct in social learning.[5]

Third, another consequence has been pointed out by Huxley (1955, pp. 8–10). The social process knows no distinction between genotype and phenotype—between germ plasma and soma. In the biological realm all information is either genetic and, therefore, transmissible through the genotype, or it is acquired by the phenotype (i.e., the organism in its life cycle) and is therefore nontransmissible. In the social realm all knowledge is acquired and, through the operation of the human creative capacity and the social codification of knowledge, it is mostly transmissible. In the biological realm there is one process for reproducing the genotype and another for maintaining it. In the social realm there is a single process that fulfills both the requirements for reproduction and for maintenance of knowledge.

Fourth, perhaps the most important consequence of the difference in the origin of the creative idea or novel behavior is the role that human values and goals come to play in the process. There is a sense in which goals play a role in both the biological and social process; in both they not only help to generate the novel idea, they also form the test of its relevance.

In discussing the biological process little ordinarily is said about the goal of the organism because it is a simple one implicit in the phenomena of life itself. The basic goal is survival. It takes the form of an instinctive drive toward self-maintenance and self-reproduction. It is this drive that is responsible for the genetic mutation and recombination that generates new ideas and behavioral possibilities. At the same time, the test of relevance of those new ideas is whether or not they serve the survival goal of the organism and are therefore maintained in the genetic pool.

In the social realm human beings develop goals that go beyond the instinctive drives associated with survival. The behavioral amplification associated with human creativity and social action makes it possible to attach value to such things as levels of productivity and levels

[5] This does not imply that a kind of information differentiation, forming a contrast to information diffusion, does not exist in the social process. This process is also different from that to be seen in phylogenesis and will be discussed in a later section (pp. 91–94).

of living. Indeed, the commitment to group behavior also gives rise to moral values. Paradoxically, these may even come to take precedence over the survival goal at the level of the organism. Thus, the social process itself gives rise to human and social goals of a different order.

But these goals still play an essential role in generating and testing novel behavior. Consider, first, their role in generating ideas. The process of human perception and interpretation has not only the capacity to perceive environmental objects and events, it is also able to order and evaluate the relationship of these objects and events in terms of the values and goals of the perceiver. The cognitive grid through which human perceptions are filtered has normative as well as logical and procedural criteria as components. When this cognitive process identifies an environmental setting that is not fully consistent with the goals, this serves to stimulate the creative process, generating new behavioral ideas calculated to modify the relationship of organism behavior and environment in ways considered to be more consistent with the goals. But the goals also form the test of relevance of the new behavior. Whether experience reveals goal convergence determines whether or not the novel behavior becomes reinforced or is apt to be replaced by still another behavioral innovation.

Fifth, it follows from this that the process of idea invention and recombination in the social realm is a much more directed process— less random and less wasteful than the phylogenetic process. It also follows from this that those ideas that generate maladaptive behavior (in the sense that they do not generate goal convergence) can be sifted from the idea pool without necessarily resulting in extinction of the biological organism. While the biological death of the organism can result from maladaptive behavior even in the social realm, the evidence of maladaptation can usually be ascertained and accommodated through adaptive behavior before this critical threshold is reached. In the biological realm maladaptive organism characteristics and maladaptive behavior could only be cleansed out of the system through death. This represents a substantial advance in efficiency. Karl Popper makes the same point in an appealing way. "Our scheme allows for the development of error-eliminating controls (warning organs like the eye; feedback mechanism) that is, controls which can eliminate errors without killing the organism; and it makes it possible, ultimately, for our hypotheses to die in our stead." (As quoted by Campbell, forthcoming.)

We might also note that, since novel ideas often take the form of changes in group behavior, it is also possible to let social forms die in our stead.

Sixth, one of the genuinely unique aspects of the social process is a consequence of the special role of goals. The ideas generated by the process of social learning fall into two classes. Most of our attention in the social sciences is focused upon those that pertain to forms of behavior and organization that are adapted to the service of human goals. But the goals, themselves, form a second class of ideas. Most of them have also been generated by the operation of the social process.

The appropriateness of novel behavior is tested by its contribution to goal convergence. If it fails that test it will usually fail to win a permanent place in the behavioral repertoire. However, the failure to generate goal convergence may not only cause the new behavioral mode to be identified as maladaptive, it may also call into question the appropriateness of the goal. In short, just as the goals form the test of adaptive behavior giving rise to the revision of behavioral ideas, new behavioral ideas sometimes form a test of the adequacy of goals and lead to goal revision. A more detailed consideration of the nature of this process will be treated later (p. 105 ff. and chap. V).

Seventh, the relationship between the environment and organism behavior is substantially modified in the social process. The biological species adapts to a state of the environment through genetic modification. The altered behavior of the species may, in turn, act to modify the environment, but this is an indirect or second-round effect not taken into account by the process of genetic modification. In the social process the environmental feedback is so direct that the relationship between behavior and environmental modification may become perceived and made an integral part of the adaptive process. Interpretation often leads to action *specifically addressed to modifying or controlling some aspect of the environment.* Sociogenesis often deliberately seeks this approach rather than always adapting behavior to the environment, as in phylogenesis.

It follows from this that social adaptation has the capacity to anticipate indirect effects of social behavior and make them direct effects through planned behavioral sequences. Sociogenesis can become future-oriented. It can be active and anticipatory as well as reactive.

Eighth, sociogenesis leads to differentiated behavior of human organisms while phylogenesis leads to homogeneous behavior of the members of a species. Some interesting consequences follow from this that we will examine shortly.

To summarize, sociogenesis, like phylogenesis, is engaged in maintaining, reproducing, and transforming environmentally relevant information that serves to guide behavior. In so doing it generates information novelties or new ideas internal to the process and subjects them to selective scrutiny by criteria (both environmental and normative) generated to a major degree internal to the process. It is, therefore, a process behaving as a creative learning system. It has the capacity to reprogram its own behavior. At the same time a description of only the major characteristics of that process reveals that it transcends in complexity, flexibility, and its directed nature the process of phylogenesis revealed by the life sciences. While we can identify certain generic characteristics, we also demonstrate that social learning exhibits distinctive features that clearly differentiate it from the stochastic learning model of biological evolution.

As we have compared social evolution with biological evolution up to this point, the contrasts have been drawn from the perspective of the socialized learning of the human organism. For the balance of this section the comparison will be continued from the perspective of social culture and the activities of organized social systems.

ADAPTIVE SPECIALIZATION AS THE ADAPTATION OF
INDEPENDENT POPULATIONS

In view of the tendency for the idea pool to diffuse and converge in the social process rather than differentiate and split, one may be justified in asking whether there is a counterpart in the social sphere to the adaptive specialization found at the species level in the phylogenetic process.

Many scholars believe that they can identify something quite similar to this in the surviving primitive independent societies and in the anthropological records of early cultural groups. Much of the work of cultural anthropologists has been directed to identifying and analyzing independent or quasi-independent culture populations. They maintain that barriers to diffusion have resulted in independent cultures arising during the passage of social history.

In the words of one anthropologist, "... through adaptive modification culture has diversified as it has filled in the variety of opportunities for human existence afforded by the earth. ... Logically as well as empirically it follows that, as the problems of survival vary, cultures accordingly change ... culture undergoes phylogenetic, adaptive development." (Sahlins and Service, 1960, pp. 23–24.) Herbert Spencer, among the earliest of the social evolutionists, had this to say: "While spreading over the earth mankind have found environments of various characters, and in each case the social life fallen into, partly determined by the social life previously led, has been determined by the influences of the new environment; so that multiplying groups tended ever to acquire differences, now major and now minor. There have arisen genera and species of societies." (In Sahlins and Service, 1960, pp. 23–24.)

It is claimed that cultural radiation can be found along the Asiatic Steppe and in the differentiation of riverine, agricultural, and pastoral societies in West Africa. It is pointed out that the early spread of neolithic culture into Europe was followed by four distinctive and widely spread cultures as the initial culture pool broke up into regional cultural variants by adapting to local conditions. (Sahlins and Service, 1960, pp. 50–51.)

In short, as man sought ways of increasing his efficiency in relating to his physical environment and meeting his material-energy requirements, he established new modes of behavior. During the earliest phase collection was the dominant mode, but it manifested itself in different specialized forms such as fishing or hunting cultures. These, in turn, became specialized under the influences of differences in biological habitat. During a later phase man learned specialized forms of cultivation or herding and these, in turn, became further differentiated under the influence of the resource base.

This type of phylogenetic branching displays many of the properties of its biological counterpart. Each improvement tends to generate social forms and functions more highly correlated with physical environment. These improvements lead to greater specialization—that is, they generate incremental changes in form and function to provide a more efficient adaptation.

A marked tendency for stabilization is also claimed. Specialization is one-sided development that tends to preclude the possibility of other changes, and that leads to stabilization. Harding makes much of this attribute of specialized culture. "The historic, archeological and ethnographic records attest to numerous instances of the persistence, 'survival,' or 'inertia,' of cultural traditions. . . . Indeed, one might formulate it as a general principle, the Principle of Stabilization, that a culture at rest tends to remain at rest." (Sahlins and Service, 1960, pp. 53–67.)

These statements indicate that something very much like the evolutionary branching of species has taken place in the social realm. At the same time this view of the social process has come under considerable criticism, even among anthropologists. It is easy to see why. Such a view tends to ignore the fact that social learning or adaptation at the level of the social system aggregate or total culture is, even in primitive social systems, a product of the socialized learning of individuals. The evolution of such systems is as much a result of the ecological coadaptation of individuals that are adapting to an endogenous social environment, as it is to the adaptation of the group to its external physical or geographical environment. Because of the influence of the social history of the group, different groups may adapt in different ways to the same physical or geographical environment. (See Benedict, 1934, and Klausner, unpublished.)

This means that the adaptation of the social group to an external physical environment is not as deterministic a correlation of social behavior with physical environment as characterized phylogenetic branching in the biological realm. Nor is the attendant stabilization so apt to terminate the history of the system. Overcoming stabilization in this kind of cultural specialization may be difficult and in fact, may not be achieved in a particular case. However, there exists in every culture (even though specialized and isolated) the potential capacity for individual human learning that may produce generalizing ideas. It may be less likely to occur than in a less specialized culture, but it is possible. Furthermore, through idea diffusion the specialized culture may have access to more general adaptive ideas from other cultures. The specialized culture may be redeemed from specializations that tend to terminate social history.

Whatever position one takes concerning the interpretation of anthropological materials, two conclusions stand out. First, while some differentiation of the idea pools of essentially independent, isolated, noncommunicating social populations may be identified in social history, the adaptation of the group to the physical environment is not as deterministic because social learning is conditioned by social history as well as by physical environment. Second, in the more advanced societies of the present-day world functional interrelationships and idea exchanges are the order of the day. The idea pools appear to be converging rather than diverging. It is not difficult to foresee the possibility, with modern communication and modern science, that cultural diffusion may lead to something like a common world culture and a virtual elimination of phylogenetic branching in the social realm as it applies to independent cultures. While this is far from realization and catastrophic events may reverse the trend, it is at least a serious prospect.

ADAPTIVE GENERALIZATION IN SOCIOGENESIS

To keep the pattern of this review intact, we should now inquire into the role of adaptive generalization in sociogenesis. As might be expected, the distinctive aspects of the social process have reinforced the importance of adaptive generalization. This is in marked contrast to the way in which new process characteristics seem to undermine the significance of adaptive specialization as the phylogenetic branching of independent cultures. Diffusion and convergence in the pool of information or acquired behavioral modes, the more direct feedback through cognition between behavior and environment, the ability to modify environment as well as organism behavior, and the more directed nature of cultural idea mutation, all tend to reinforce the process of adaptive generalization.

Thus, as we trace the social process through the successive dominant cultures of early societies and through the phases of cultural convergence that seem to characterize the Western world today, we can note a series of opening improvements that has led to greater environmental independence and control, and has generated more idea-rich, behaviorally complex forms of culture. These generalizing or progressive lines can be traced through every aspect of social behavior.

The artifacts that amplify the power of human behavior display a progressive increase in their complexity and idea content. They now represent embodiments of mind and muscle that rival man in many of his activities. Their functional purpose also displays a significant progression. Early forms (the club, the wheel, the trained beast of burden) were designed primarily as extensions of man's limbs and supplements to his muscle. The more dramatic of the recent forms (the computer, electronic communications, modern transportation modes) appear to be primarily extensions of man's mind (the externalization and enlargement of both logical capacity and memory capacity) or are directed to the more efficient coordination of control of an increasingly complex culture.

Similarly, the evolution of man's social groupings or social systems stretches from the family, tribe, and guild in earlier phases to the modern industrial complex, the urban complex, the nation state, and international coalitions in current phases.

The cognitive processes of mankind display a similar line from the language of primitive referents to formal codification of knowledge, to mathematical and symbolic logic, to advances in epistemological understanding, to the point where a significant portion of man's energy is addressed to trying to understand and consciously direct and control the creative process. We have "invented the method of invention" as A. N. Whitehead once remarked.

Throughout this process we observe the same general characteristics of adaptive generalization that we have seen in phylogenesis. As before, the generalizing adaptations are identified with major new social forms and functions. As before, the major generalizing adaptations tend to survive in subsequent cultures.[6] As before, the generaliz-

---

[6] This may be illustrated by the following example: As man emerged from the presocial phase he met his material needs by employing his superior mobility and understanding of the environment and his primitive socialization to collect from the natural environment those resources he needed. As he progressed he came to effect a concentrated rearrangement of environment by establishing agriculture and animal culture. As he advanced still further he generated more complex rearrangements of environment through fabrication. But these were not discrete phases. Each later phase was grafted on to the roots of the older and in the most modern culture the elements of all three basic processes remain intact and are, indeed, essential components of a complex system.

ing adaptations display the characteristic increases in organization and adaptability.

Increased Organization: Adaptive Specialization in Social Subsystems. At the same time that phylogenetic branching appears to be disappearing as an attribute of separate independent cultures (corresponding more or less to independent biological species), specialization has reappeared, in the form of specialized social subsystems. While specialized behavior is declining in sociocultural systems in the large, it has become more prevalent in their components. It is as if a process inversion were taking place.

As man's knowledge was expanded and came to be embodied in his tools, the sum total of his acquired knowledge became so great that individuals found that they could only increase the idea content of one function or activity by reducing the number of functions performed. This could be done only by transferring some essential activities to other individuals, thus permitting an expansion of the energy and mind applied to the activity. The activity externalized by one organism must, if an essential function, be internalized by another and, in turn, be accorded a larger application of energy and mind. Accordingly, the individual comes deliberately to take on different components of the total of mankind's acquired behavior.

As the individual specializes and therefore externalizes essential functions, there occurs an extension and complication of his lines of association with and dependence upon other individuals and the implements, institutions, and ideas they employ. Ideas have to be communicated, specialized outputs transferred and transported, and the entire evolving network of interrelated activities subjected to some regulation and control. Specialized integrating activities (e.g., business and public management) and linking activities (e.g., transportation and communication) emerge and bring functionally related activities into more closely managed or more directly linked sets, giving rise to functionally related activity subsystems within the total. Combinations of individual specializations are incorporated into broader functional specializations and thus give rise to new social forms (e.g., the enterprise, the city, the nation, etc.). Indeed, those organizational ideas that have permitted the specialization of human adaptive behavior and the integration of

91

these specializations into behavioral social systems are among the most significant adaptive generalizations of the cultural sphere.

To take an example, in a modern sociocultural system agriculture does not appear as a divergent cultural form as it did in an early period of social history. Rather it appears as a specialized subsystem component of a complex social system. The total social system has established boundaries forming a subsystem agricultural specialization.

Furthermore, advancing technology and social organization expand the number of feasible partitions of specialized agricultural activity. The number of micro-habitats that support further activity specialization is enlarged. Thus, agriculture becomes differentiated into specialized field crop, fruit, vegetable, and livestock producing subsystems. Livestock systems may become further differentiated into dairy, meat livestock, and animal fiber systems. Meat livestock becomes further differentiated into specialized cattle, hog, or poultry systems. The total agricultural subsystem also develops further partitioned activity specializations, as cotton ginning, corn milling, hay bailing, animal doctoring, poultry hatching, etc., move into specialized activity niches of their own.

Something like this happens to many other components of the social system. In our complex modern cultures the great variety of this efflorescence of specialized activity may commence to rival the variety of specialized biological forms in nature. In this form adaptive specialization is still persistent.

Thus, we can see that most of the traditional characteristics of adaptive generalization in phylogenesis are matched by similar characteristics in the social process. There is a marked *increase in the complexity* of the total social organization. This is accompanied by *increased specialization* of subsystem components. This, in turn, has led to the organization of these specializations into higher levels of system integration that have become progressively more general and inclusive as one moves up the system hierarchy. There is, therefore, a perceptible movement towards *centralization*.

At the same time we can also see that while these generic characteristics of adaptive generalization can be observed in the social process, there they are manifest at system levels different from those of phylogenesis and so do not form a perfect analogue. The variations stem from the differences between the two processes.

First, adaptive generalization at the species level in phylogenesis produces a population of organisms with roughly similar generalized behavioral capabilities based upon common attributes of an interbreeding genetic pool. Adaptive generalization in sociogenesis produces a population of human organisms which show quite different behaviors. Phylogenesis tends to reduce the statistical variation between organisms in the information they share with the genetic pool. Sociogenesis tends to increase the statistical variation between human organisms in their share of the information in the cultural pool.

Second, in the biological realm adaptive generalization leads to increased complexity in the organization of the organism. This is manifest in increased specialization of the component organs and their increased subordination to centralized control. Increased centralization also seems to attend increased social organization. Since human activities are differentiated as a result of idea accumulation and subsystem specialization, the coordination of these linked activities becomes essential. This has led to a social system hierarchy where certain aspects of subsystem behavior are subordinated to higher system control.

However, this subordination of parts and centralization of control are much more restricted phenomena in the social realm. Unlike the subsystem components of the biological organism, the smallest social subsystem (man) is characterized by a creative capacity and a decision-making and self-activating capacity of wide and flexible range. In short, the creative and decision-making attributes of the social process are present at all system levels. Thus, higher order systems often leave substantial elements of control and considerable behavioral independence to lower order systems.

In contrast, the most advanced biological organisms have gathered the creative activating capacities of the organism into a single node. Where decision options are reserved to its component organs (e.g., autonomic responses) they are of a mechanistic and primitive sort that are not characterized by adaptive flexibility and do not rest upon conscious adaptations.

Thus, organization does occur in social systems and some degree of centralization of creative and decision-making processes develops. But the decision-making and creative faculties are not all drawn to one node. Indeed, they ultimately remain embodied in the smallest component of culture—man. Such centralization as occurs is developed by many individuals each of whom exercises his creative faculties with others to bring about cooperative social systems.

Third, by the same token the behavioral specialization of the human organism that accompanies social organization need not constrain and narrow the behavioral range of the human being. This is true partly because man often performs multiple roles and participates in the behavior of different social subsystems. Even if his behavior with respect to each is specialized, his overall behavior may be quite diverse. More important, while increased specialization and diminished behavioral range may attend occupational specialization as a component of social organization, the very act of specialization may yield for the individual a larger share in a larger and more variegated social product. This can lead to more generalized behavior in consumption activities supporting a wider variety of behavior and experience.

Increased Adaptability. Adaptive generalization in the social realm is marked by increased adaptability as well as by increased organization. Indeed we have seen that one of the two principal features of social learning is the phenomenon of socialized human learning at the level of the human organism. Social learning is the product of that unmatched adaptive generalization of phylogenesis—the supremely adaptable human organism. That adaptability is the attribute out of which sociogenesis emerged. In the process, the phylogenetically created adaptability of the human organism has been further amplified by the social process itself.

For example, the characteristic of adaptive generalization in phylogenesis was a progressively improved faculty of discrimination. This was supported by improvements in the efficiency of environmental receptors and the emergence of a primitive analytical capacity. We have already seen how in man these trends were brought to the critical point where codification and transmission of acquired behavior became feasible. The faculty of discrimination is further raised to a new level by the social process. The powerful amplification of man's biological receptors through the use of material implements, the extensive and progressive development of the modes of knowing and the development of implements to support and amplify the memory and logical power of the human organism are all examples of ways in which the social process has amplified the discriminating capacity of the human organism.

A second characteristic of adaptive generalization in phylogenesis was a marked increase in mobility which further amplified man's discriminating power by expanding his environmental range and developing further his discriminating behavior. The social phase is marked by this same trend. As social development has progressed, social activity has been marked by increases in mobility to the point that it can extend its influence into the most remote and inhospitable habitats of the earth and project its presence into space.

The improved faculty of discrimination and improvements in mobility permitted emergent biological forms to develop a wider range of environmental tolerance—partly by improving the versatility of response and partly by bringing ranges of environmental variation under direct organic control. There can be no doubt that both of these traits stand out in social evolution. Particularly significant is the way in which man has come, in advanced cultures, to live in an essentially man-made environment. Mankind in these advanced settings has won considerable freedom from the constraints of physical environment in the raw. For the human organism the most serious remaining environmental challenges are associated with social environment. It is his social environment that now seems more often to evade effective understanding and control, and the process of cognition is turning inward to the analysis of culture, social organization, and the psychic attributes of man himself.

Just as in the biological realm we came to recognize that the adaptive generalizations leading to improvements in the nervous system were of critical importance, we recognize here the critical significance of improvements in the psychic attributes of culture. Indeed, we identify the social process as a progressive increase in the development, codification, and embodiment of new ideas leading to and derived from more complex social organization and greater social adaptability. The development of oral, then written language; the development of mathematics, logic, and the scientific method; the development of artifacts to extend the power of the human nervous system; the development of all the artifacts that support life, the reproduction of culture, and the creative process itself; the development of functional groupings of human activities by the specialization of production and consumption; and the emergent forms of government and management that seek to coordinate the decision-making and creative potentialities of mankind are all supported by a progressive increase in the idea content of culture.

In short, it is precisely those attributes of adaptability marking the birth of sociogenesis that have been further augmented by the social process itself. An essential instrument of this process has been the emergence of social organization and social systems that integrate specialized human behavior. This calls our attention to another way in which adaptability has been enhanced by the social process. In the preceding paragraphs we have emphasized the way in which the social process has amplified adaptability at the organism level of social learning—i.e., the adaptability of the human being. It has also done so at the social system level.

Social systems are inherently more adaptable than biological organisms in one important respect. The organization and behavior of human social systems can be modified on a part-by-part basis while still maintaining the viability and continuity of the social system. This is a trait that cannot be matched by the organization of the cells in the body or even the social organizations of insects. (See Campbell, 1965, p. 30.) This attribute gives to social systems a measure of adaptability not matched by any prehuman biological behavioral systems.

Furthermore, as social evolution increases the complexity of social systems and the variety of component behaviors, the adaptive flexibility of the system is enhanced. There is considerable evidence that those systems that internalize a wider variety of behavior are more adaptable than the more specialized. At the level of the urban system, for example, a Boston is more resilient than a Pittsburgh in the face of social change. At another level, a large multiproduct business corporation is more adaptable than a single-product private producer. Indeed, man has invented legal forms of social organization (e.g., the corporation) and has developed habits of behavioral integration through managed ecosystems (e.g., nation and urban systems) that facilitate the process of part-by-part reconstruction.[7]

[7] The foregoing comparison of adaptive specialization and adaptive generalization, as they are shown in the biological and social processes, is lacking a discussion of

EVOLUTIONARY THRESHOLDS AND EVOLUTIONARY ACCELERATION

One of the striking characteristics of the process of biological evolution, when observed at the level of the biosystem as a whole, is the fact that several major thresholds have occurred where the nature of the evolutionary process itself is observed to change. The first of these, of course, is the process change that initiated biological evolution itself. Oparin (1938) has described the process of chemical evolution whereby the original gaseous mass that was to become earth separated from the sun and evolved into a complex array of chemical compounds (including the hydrocarbons and nitrogenous compounds that were a necessary precondition for life). These passed through stages leading to colloidal systems with the power of chemical assimilation. Out of this colloidal process evolved material organization with assimilative and reproductive processes that, through a phylogenetic process at the population level, evolved a series of elementary life organisms. But this phylogenetic process was different in character from the process of chemical evolution.

Another major process threshold can be identified with the fact that, in time, the evolution of biological populations at the species level came to be more prominently influenced by the biological environment than by the geological environment. The process was characterized by a kind of biosystem-level coadaptation, and the evolution of the biosystem itself generated new adaptive zones that encouraged adaptive generalization and accentuated progressive trends.

A third major threshold is associated with the fact that phylogenesis created improvements in organism adaptability until it generated learning organisms. Such an organism, we have observed, has the power to modify its own behavior during its life cycle without sole recourse to another round of selective transformation of the genotype.

Each of these thresholds had the effect of modifying the way in which the evolutionary learning process was itself carried out.

---

two other topics developed in the articulation of the biological process, namely, the preconditions for adaptive generalization and the relationship between adaptive generalization and adaptive specialization. These concepts have characteristics that are generic to both processes, but the nature of the similarities will come into sharper focus in later chapters, after the concept of social learning has been more fully developed.

In a similar way several process thresholds are observable in social evolution. The first of these are those changes that permitted learning organisms to share individual learning experiences that led to the emergence of socialized organism learning. In the early phases of social evolution human learning was dominated by the sharing of information and behaviors having to do with the individual's relationship to the nonhuman environment. The adaptations sometimes spread rapidly because of the demonstration effect and human communication, but the adaptive behavior was not, itself, highly integrated or socialized in nature.

A second threshold emerged as the social organization of behavior addressed to social maintenance progressed. The organization of the physical transfers and transformation that mark the everyday task of maintaining human populations gave rise to social systems, communication tendrils (along with their symbols of language and numbers), and transportation linkages. Man became progressively immersed in a social environment made up of the behaviors and responses of other individuals and social groups. The behavioral amplification associated with social organization permitted improvements in efficiency which, in turn, permitted improvements in the levels of biological and physical welfare and/or increases in the sizes of supportable populations.

Still, during this phase, the adaptations of individuals and social groups were primarily reactive to specific environmental problems. Taken together, however, these adaptations represented a form of coadaptation not altogether different from the coadaptation that emerged when biological evolution in its second phase began to adapt to biological environments. This feedback between behavioral change leading to modifications in social environments leading to further behavioral change, when combined with the direct diffusion of ideas, led to a substantial progressive accumulation of new ideas and behaviors and to progressive levels of social advance. Most of human history evolved under the influence of a social learning process that operated essentially in this way.

In the last few hundred years we have crossed a third major threshold. In the social process perception, interpretation, modified action, testing through achievement of goals, reinterpretation, etc. operated for many hundreds of years before it was perceived and understood as a total learning process. In time it was recognized that the process could be consciously applied, not bit by bit as components in a reactive adaptation, but as a total process in an active anticipatory adaptation. Scientific method came to be consciously applied to the creation of new ideas and the testing of their validity.

This was a change in the process of social learning comparable to the turning point in biological evolution where the adaptive transformation of the genotype was progressively supplemented by the behavioral adaptation of the phenotype. At this stage of the social process the reactive adaptation of the individual to the problems presented by a social and biological environment comes to be supplemented by an active anticipatory adaptation understood as a total learning process. It is a process that can be used to actively search environmental relationships independent of individual, pragmatic, short-run problem solutions. Applied as a total process, social learning can be directed to the *shaping of behaviorally patterned environments,* in contrast to the slower emergence of *environmentally patterned behavior* more common to history. Learning processes can become more explicitly future-oriented and more efficiently directed toward the understanding and control of external higher order environmental systems. Social learning is emerging as a process that can advance through the application of a learning technology and without *sole recourse to the reactive individual adaptations indirectly related through environmental feedback.*

However, the conscious understanding, articulation, and application of the process of social learning are still largely dominated by the application of formal scientific method to the understanding and control of man's nonhuman environment. In the last few centuries physical science and technology have made tremendous strides in revolutionizing the life of man by increasing physical productivity and human welfare and by generating vastly more complex social systems essential for the application of modern techniques. But we have become increasingly conscious that the concept models generated by physical science and the very process embodied in the classical scientific method are not adequate to permit mankind to cross still a fourth threshold. What is needed is the ability to come to a conscious understanding, articulation, and application of the process of social learning as it operates to transform social or human environments. The social and behavioral sciences have placed a foot on this threshold in recent years, but the threshold is yet to be crossed.

Associated with the progressive thresholds in evolution is another feature of considerable interest and importance. Each threshold is associated with a change in the way in which the evolutionary process, or learning process, operates. These changes have had the effect of changing the efficiency of the learning process and are, therefore, associated with an acceleration of the evolutionary process. A few figures illustrate this point.

The total span of the chemical evolution that generated our planet and led to the threshold of life cannot be accurately known, but the best estimates deal with an order of magnitude of 10^9, or billions of years. In the biological phase major changes such as the evolution of the fully specialized horse, or the evolution of birds from reptiles, required a time order of magnitude in millions of years or 10^6. As we move into the social phase, major cultural change in the lower Paleolithic era required something on the order of a million years. By the upper Paleolithic era the period was nearer ten thousand years (10^3). During the prescientific historical era the time came down to 100 years and, since the advent of science, each decade during the last century has seen at least one major change. Today even the most momentous changes, such as the discovery and first practical application of atomic fission, may take only half a decade. We can see that the order of magnitude of time involved in major change has dropped progressively from nine to less than one.[8]

This acceleration of the learning process makes it imperative that we learn to understand the process of social learning as it works itself out in social environments so that we can make sure that the process itself conforms to human objectives. A suspicion is emerging that the uncontrolled process may be leading us into some form of evolutionary trap that will place the human species itself in jeopardy.

EVOLUTIONARY TRAPS

We are aware that the adaptive specializations of species in phylogenesis often lead to biotypes whose adaptive behavior and genetic

[8] This paragraph is a paraphrase of a similar point made by Huxley (1955).

variability are insufficient to permit survival in the face of large-scale environmental change. These evolutionary traps are a well-known part of the story of biological evolution. This leads to a significant question about sociogenesis: Does behavioral specialization in the sociocultural sphere hold a similar threat?

The answer is that it may, but that it tends to be considerably less deterministic in effect. The issue is a more complex one in the sociocultural realm. Consider, first, the process of adaptation at the level of the individual in society.

Let us remind ourselves that in phylogenesis an evolutionary trap occurs when a behaviorally specialized individual or a population of individuals fails to modify its behavior in a manner appropriate to an altered environment. This maladaptation leads to individual or collective extinction. Similarly, occupational specialization of the kind yielded by social organization can leave an individual—or some homogeneous set of specialized individuals like blacksmiths, bookbinders, pensmen, or potters—high and dry when a changing social environment makes his specialization obsolete. There is no doubt that these specialized subsystems have often established a considerable degree of cultural isolation while still retaining some external ties, and that these semi-isolated specializations can be conservative and resistant to change. Social history abounds with records of broken men who could not escape a specialized activity niche in the face of radical social change. However, for several reasons the failure of the individual to at least survive is not so deterministically assured in this case.

First, the individual's behavior is not transformed through discrete genetic transformations that require physical death and cohort sequences to effect. It is transformed through acquired behavior—and by the individual and social learning process. Having some internal capacity for learning, the individual can often effect adaptations that permit at least a partial escape from the trap of specialization.

Second, as noted earlier (p. 94), the specialized individual often retains a range of knowledge and experience that exceeds the information requirements of his specialization.

Third, the individual frequently has access to learning resources external to himself. Specialized educational and training functions may exist to which he can turn for assistance in behavioral transformation and social adaptation. Indeed, in an ethical society there is an additional ameliorating factor. The disadvantaged individual may continue to share in the joint output through social transfers after an obsolescent specialization terminates his productive contribution. He not only has internal and external access to learning capacity, but he may also be subsidized by the system during the period essential to his adaptation.[9]

In this connection it is worthwhile noting the existence of a persistent but mistaken idea about the nature of the evolutionary trap in the social process. Classical social Darwinism has long raised the specter of one trap that turns out to be false. This view maintained that diffusion and convergence in the cultural sphere should be resisted and attacked on the grounds that welfare requires the competition of diverse cultures. Such an argument has been employed, for example, as a justification for maintaining class structure and inequality of opportunity. It seems to be based upon improper analogies with biological evolution and the assumption that only in this way can cultural (like genetic) variety be maintained. It is claimed that cultural convergence will lead to a homogenized culture that will lose its vitality. The implication is that society will make a vast closing adaptation. But, as the discussion above has shown, this argument does not take into account the efflorescence revealed by adaptive specialization in the form of subsystems. (See Dobzhansky, 1962, pp. 322–25.) Indeed, it is now recognized that there is an offsetting advantage (if not an ameliorating effect) in the nature of occupational specialization itself. There is statistical variation in the genetic base of the biological organism and attendant organism behavior in the human species as in all species. Since activity specialization provides a wider range of environmental niches in terms of the productive (creative) activities of man, genetic and behavioral traits that could not survive under the environmental conditions that characterize most biological species now find a viable environment. There also exists a wider range for human creativity and individual expression that often contributes to the self-realization of the individual.

While individuals may be confronted with evolutionary traps by the

---

[9] The form of subsidization is important. If not properly designed the subsidy may actually inhibit social adaptation: witness some aspects of the agricultural adjustment policies of the United States.

social process, we have to conclude that their effect is less deterministic than in the biological realm. Physical extinction has usually been transmuted into existential suffering (a relativistic rather than an absolute consequence).

Now consider the question of evolutionary traps at the level of the social system. The survival characteristics of social systems, as such, are also remarkably enhanced by the social process. First, they are also the beneficiaries of the adaptability that characterizes the socialized learning of the individual. Second, social systems come into being to generalize and coordinate specialized individual and subsystem behaviors; as an entity, therefore, they tend to encompass a broader behavioral range. Third, as has been pointed out, social systems can be modified on a part-by-part basis opening a way to continuity of the system in the face of change. Even if, as sometimes happens, a group behavioral system fails to survive as a system, individual or group subsystem components may survive by attaching themselves to other social systems that utilize similar behavioral capacities.

All of these factors lead to an interesting contrast between biological and social processes. In phylogenesis the species may survive though the individual dies. In sociogenesis the individual may survive though the mode of specialized behavior or the social subsystem of which he is a part may become obsolete.

This is not the end of the issue, however. The idea of the evolutionary trap presents itself in a new guise in the social realm. We are all familiar with the way in which complex high-order social systems like empires, nations, and states, have flourished for a while in social history and then disappeared. But these are not examples of radiating cultures that become specialized, isolated, and obsolete. They often represent the highest forms of social organization and function characteristic of their era, and they have not always disappeared because of the competition of a stronger social system. There is a source of social decline unique to sociogenesis.

The success of a social system is not solely a result of the efficiency of its interrelated specializations in fulfilling the needs for self-maintenance and reproduction of its population. It is not even related solely to its ability to provide higher per-capita incomes to its constituents.

All social systems are behavioral coalitions based upon consent. The consent of constituent behaviors may be gained by compulsion based upon control of power by a few, by the distribution of a reward based upon the increased social products associated with the behavioral amplification made possible by social action, and by the psychic internalization of norms associated with either the authority of power or the advantages of cooperation.

Social systems based upon compulsion are always liable to disruption because social change has a way of disrupting the balance of power. Voluntary coalitions also are subject to deterioration because they are sometimes not able to reconcile conflicting and multiple goals of the constituents. Internalized norms break down under stress. The hardest problem for sociogenesis to solve is how to practice the technique of increased organization and subsystem specialization with its attendant benefits in a context that fully integrates the vital creative resources, preserves the dignity, and promotes the willing participation of its human components. The evolution of social values and goals has not progressed to the point where we have succeeded in devising the integrative social system that fully reconciles the objectives of its constituents. At best they satisfy for a while, but alterations in power positions or the modification of goal structures resulting from social change may from time to time dissolve the coalition of consent. The social system may suffer a degree of regression or disintegration, not because it is excessively specialized or trapped by the force of external events, but simply because the internal glue of its value structures or power relationships has become unstuck. Such an event may constitute a very real evolutionary trap, not only for the system but for many of its constituent subsystem populations and individuals.

There is another potential evolutionary trap unique to the social realm. Man's adaptive ability, extended through social action, has enabled him to operate successfully and deliberately in transforming the environment. The potentials of some of these environmental changes are so substantial that misuse could constitute a threat to survival of the human species.

We are confronted with the reality of atomic power. Equally significant is the possibility that human cells can be programmed with synthetic messages within twenty-five years (Nirenberg, 1967). In short, social learning has advanced to a point where we stand on the threshold of bringing the process of phylogenesis that spawned sociogenesis under the direct control of sociogenesis. The parent process is coming under the control of the offspring. Furthermore, we are just becoming aware that the impact of a large human population utilizing advanced technology for environmental control and exploitation may be upsetting the biological-geological-climatological ecology of the world with serious consequences for the future.

The threat that these developments offer arises out of the fact that they are the product of the successful organization of the process of social learning as it applies to the understanding and control of man's nonhuman environment at a time when the organization and control of social learning, as it applies to man's social environment, remain incomplete. The progress of natural science remains uncontrolled and unconstrained by a mature human science. These developments subject mankind to the potential of an especially worrisome kind of evolutionary trap.

The Unique Attributes of Social Behavior

Having compared the process of social evolution with the process of biological evolution, we are aware that the learning system we know as phylogenesis displays a number of characteristics that seem to be generic to creative learning systems, and that these attributes can be observed at work in the social process as well. At the same time, we have observed that the social process differs not only because the generic traits manifest themselves in different ways and at different levels of behavior, but also because a number of characteristics of the social process are unique to it.

The most distinctive thing about human social behavior is that it is based upon a conscious cognitive process which displays several activity components. Human social behavior is directed to organizing behavior and analyzing behavior. At times both of these component behaviors are exercised in the service of behavior directed to changing behavior. In this section we shall explore the nature of these unique attributes of social behavior.

BEHAVIOR DIRECTED TO ORGANIZING BEHAVIOR

Behavioral specialization, which in man is his response to a continuous accumulation of knowledge, is not in itself unique to sociogenesis. Phylogenesis has, on occasion, created species that take on behavioral specialization and social organization. The insect societies are a good example. Their specialization into such roles as scouts, workers, and soldiers in an interdependent economy is well known. Furthermore, this social organization displays a selective advantage in phylogenesis for a reason similar to one of the advantages of sociocultural societies. Through specialization there are operating efficiencies to be enjoyed in exploiting food supplies (material and energy resources) that are transportable and storable.

At the same time, there is an important difference. The behavioral specializations of the insect society are phylogenetically evolved and genetically fixed; their functional relationships are rigid and each society is identical with every other society of the same species. The social systems generated by sociogenesis, on the other hand, grow out of adaptive behavior and are not identical duplications of a rigid genetic templet; they are formed to meet the requirements of a unique environmental situation.

This gives rise to a new role for behavioral specialization in the social process. We learned earlier that human specialization developed because only through specialization and group cooperation could man embody in useful behavior all of the accumulated acquired knowledge. Only in this way could he gain access to the behavioral amplification that shared knowledge and shared behavior could bring. Since these specialized behaviors of human individuals and groups of individuals are not coordinated and linked functionally by fixed and instinctive patterns, action must be directed to consciously performing these functions. This leads to additional specializations devoted to organizing behavior, which take two major forms: (1) those that link the specialized behaviors through transfers, and (2) those that perform a coordinating or management function.

With reference to the first, the specialization of functions and the consequent emergence of social systems greatly increase the inter-organism and intersystem transfers of signals, ideas, and material artifacts. Specialized activities require information and material and energy inputs from other specialized activities. Since their output is specialized, it must be exchanged with other specializations to gain access to the advantages created by behavioral amplification through specialization. The accumulation of socially acquired behaviors and the specializations they entail do not have to proceed very far before the exchanges of information and goods can no longer be carried on in conjunction with the specialized production or transformation activity. The multiple channels of exchange required in a complex social system, and the energy required to carry out the exchanges, advance to levels where they must be separated from specialized production activities. Accordingly, transportation and communication specializations arise. They are essential to organized social behavior.[10]

With reference to the second form, the facilitation of transfers does not assure that the inputs and outputs of each system match to form balanced flows adequate to sustain the social process. Consequently, there is a role for a management function to receive signals from the input-output termini of the system and to match the constituent behaviors and their resulting flows of products and services in response to these signals. This behavior-organizing behavior is unique to the social process. Where these transfer flows and their associated transformation or production activities are not matched by a direct control procedure, the social process has innovated such informal institutional controls as markets.

Behavior-organizing behavior not only has the task of creating a viable balance, it must also act to maintain and restore the balance in the face of change. No social system is completely isolated from external influences. Exogenous change may dictate changes in the

---

[10] Note the differences between the improvements in mobility resulting from adaptive generalization in sociogenesis and those generated by phylogenesis. Improvements in mobility in phylogenesis mean an improvement in organism mobility primarily. Improvements in mobility in the social realm may occasionally mean that some social subsystem becomes in a sense more mobile as an entity, and it undoubtedly has the effect of increasing the mobility of the human organism through the invention of transfer amplifiers. However, they have especially served the interests of improved efficiency in social organization. It is a necessary functional corollary to behavioral specialization.

107

organization and utilization of existing behavioral specializations; it may also lead to forms of reorganization that can be characterized as behavioral development.

BEHAVIOR DIRECTED TO ANALYZING BEHAVIOR

The organization of specialized human behavior into integrated social system behavior implies a social purpose or a value to be derived from shared behavior, and it implies a means of maintaining the behavior under the control of the group goal. Both of these are aspects of one of the unique characteristics of social behavior, namely, behavior directed to analyzing behavior.

The existence of such analytical behavior is, of course, a manifestation of the purposive, conscious character of human cognition. We observed earlier that the behavior of animal organisms can be purposive under the motive of instinctual drives. We also observed that this fact could not impart a purposive character to phylogenesis as a creative learning process at the level of the species or population.

Purposive behavior in human organisms is of a different character. In man the cognitive process advanced to the point where the generalizing, abstracting, symbolizing power was sufficiently great for man's creative imagination to generate the concept of the environment. As soon as this concept was created, man ceased to live in natural and innocent communication with this environment. He created a conceptual boundary in the continuum of nature and thus created the concept of self—a bounded entity set apart from environment—and a new mode of cognition. Cognition becomes self-conscious. It knows that it knows and commences to have some insight into the manner of its knowing.[11]

It also has the capacity to develop values and goals that are not based upon instincts, but are derived from the context of social action.

---

[11] The use of the term "self" has been rejected in some quarters as unmeaningful from a scientific point of view. However, the term is not used here to suggest the concept of a self (or a soul) that performs acts, solves problems, or steers conduct in a trans-psychological manner not accessible to objective analysis. The term is used in the same sense that psychologists like Allport use it to denote the "propriate" functions of personality. (See Allport, 1955, pp. 54–55.)

Unlike the biological process, where purposive instincts of the organism cannot impart purposive behavior upon phylogenesis operating at the level of the species, the purposive nature of human activity that grows out of human cognitive capabilities can be generalized to the behavior of social systems at the level of human populations. Almost every social system or manifestation of group behavior is based upon a general social purpose—the encouragement and facilitation of the amplification of some modes of human behavior. The specific goals are associated with the specific nature of the task (e.g., increased productivity in agriculture, increased mobility through organized transport, the sharing of creative experience through a symphony). In each case there emerges a whole set of subsidiary goals or instrumental efficiency criteria that act in the service of the more general system goals.

Social organization is motivated, then, by these human value payoffs. In making them effective in group action it is necessary to apply the human cognitive process at the social system level. First, the behavior of the system needs to be monitored. This requires the collection and examination of evidence relevant to control monitoring. This is akin to perception at the organism level and, like the basic cognitive process, the evidence that is selected for examination out of the almost limitless array of exogenous and endogenous objects and events is a function of its social purpose. Perception is partly under the control of normative criteria.

Second, monitoring the behavior of the system requires not only the collection and examination of evidence (social system perception), but also interpretation of the evidence. This is the testing or analysis of the evidence of system performance in relationship to the system goals. These modes of behavior directed to analyzing behavior are essential to social system viability and control.

The same analytical capabilities may also be engaged in creating and testing behavioral options in the service of system goals, or in evaluating and revising the system goals themselves.

Behavior directed to analyzing behavior and behavior directed to organizing behavior are, of course, not independent, but are closely interrelated attributes of social behavior. The active process of social organization and control is guided by analysis, and analysis is molded by the goals and problems associated with active social organization.

BEHAVIOR DIRECTED TO CHANGING BEHAVIOR

The attribute of social behavior that is most significant for our purpose is the behavior directed to changing behavior. It is this attribute that gives the distinctive character to social learning. It adds a new dimension to behavior directed to organizing and analyzing behavior.

Behavior Directed to Social Maintenance. First, it is important to point out that social learning is not a necessary consequence of all social behavior. Perhaps a major portion of behavior-analyzing and behavior-organizing behavior is directed to social maintenance. It is homeostatic in character. Once a social system is formed, most of the action that takes place under its control is directed to carrying out the purpose and activity of the behavioral potential that defines the system.

The social entity may behave much of the time in the fashion of the machine system open to resource exchange, described in chapter I, or like a lower order animal organism that reacts to its operational environment in the process of seeking out the thermodynamic throughput (food, etc.) that maintains it. Such a behavioral system may alter its manifest behavior as it encounters environmental changes at its input or output termini, but it does so within the context of an established behavioral mode and system objective.

Thus, social behavior may be absorbed with (1) the task of collecting and examining the evidence that reveals the states of the operating environment at input-output termini and that maintains the organization of endogenous component behaviors, (2) interpreting the meaning of these signals in the light of system objectives, and (3) managing the component behaviors so as to reconcile the system objectives in the most efficient way with the operational environment. Such social behavior is absorbed in the service of maintaining a steady state system— a system whose throughput cycle operates continuously to carry out the prescribed input-output mode.

Behavior-changing Behavior. But all social behavior is not reactive and absorbed with maintaining behavior. The human cognitive process has the capacity to be active and anticipatory in nature; to promote

its goals through behavioral innovations; and to discover new knowledge and new modes of behavior through "play"—that is, through
behavior that seeks understanding and experience not immediately
relevant to self-maintenance. It has the capacity to modify old goals
and develop new. These creative capacities are also displayed in the
behavior of social systems. These sometimes encounter changes in their
operational environments that create major discrepancies between objectives and behavior—discrepancies so large that routine social analysis and social management cannot satisfy system goals by reorganizing
traditional behavioral modes. The failure to satisfy the goals of the
system constitutes a problem that stimulates the creative attributes of
behavior-analyzing and -organizing behavior. New kinds of evidence
are sought about the behavior of both external and internal environments. This evidence is examined for clues to the nature of the problem and to modifications in behavioral mode that may better satisfy
the goals of the system.

Behavior-organizing and behavior-analyzing behavior may be pressed
into the service of designing and implementing new behavioral modes.
Social systems may even devote these forms of behavior to creative
"play"—to the environmental search and behavioral design that are
directed not so much to the solution of immediate problems as to the
identification of new opportunities for satisfying human goals. This
problem-solving, opportunity-seeking attribute of the human cognitive
process gives rise to behavior directed to changing behavior—a higher
order attribute of human cognition and the social process. Behavior-
changing behavior must, of necessity, subsume the other two attributes
of behavior, but, as we have seen, behavior-analyzing and behavior-
organizing behavior may be carried out independently of the former.

Social Learning: A Process of Evolutionary Experimentation. All of
the discussion of the last few pages, dealing with the unique attributes
of social behavior, leads us inevitably to our central thesis: that there
is a process of behavioral development or social learning implicit in
behavior directed to changing behavior. Our task now is to understand the nature of this process and its implications for development
planning and social problem solving.

We already know that this process differs from the learning process that is phylogenetic evolution. It is consciously practiced; it is socialized through shared ideas and action; and it is purposive, directed, and goal-oriented. We must now find out whether there are rules that govern its conduct and provide a guide to the practice of behavior-changing behavior.

The chapters that follow will explicitly address this question. The position to be taken is that there are such rules; that man has already identified and tested some of them in the form of the procedural rules of classical science, but that this is a limited subset of the characteristics of the learning process fully applicable only to certain problems having to do with man's nonhuman environment. The procedures of classical science, we shall maintain, are a special case of a more general process characterized here as evolutionary experimentation.

Part III

SOCIAL LEARNING AS
EVOLUTIONARY
EXPERIMENTATION

IV   The Nature of Prediction and
      Planning in Social Learning

IT WILL BE RECALLED that in chapter I we examined various concept models that have been applied to the analysis of social change, and we identified a set of growth models and a set of development models viewed as programmed learning. Each organizing concept implied a different approach to the prediction and planning associated with social action. In the present chapter the base provided by these concepts will be used to examine the approach to prediction and planning that is implied by the idea of a creative social learning process. Such an exercise may yield for us our clearest understanding of the nature of the process, for it is through prediction and planning that the process is revealed in action.

Two Modes of Prediction and Planning in Conventional Science

In conventional practice and conventional science there are two kinds of prediction and planning. Popper (1964, p. 43) makes this distinction: "In the one case we are told about an event which we can do nothing to prevent. I shall call such a prediction a *prophecy*. Its practical value lies in our being warned of the predicted event, so that we can ... meet it prepared. Opposed to these are predictions of the second kind which we can describe as *technological predictions*. ... They are, so to speak, constructive, intimating the steps open to us *if* we want to achieve certain results." For example, "We may predict (a) the coming of a typhoon, a prediction that may be of the greatest practical value because it may enable people to take shelter in time; but we may also predict (b) that if a certain shelter is to stand up to a typhoon, it must be constructed in a certain way."

Both modes of adaptation appeared early in the process of social evolution in limited applications. For example, over the time span relevant for human adaptation, the cosmological forces operating on man's environment are essentially machine-like in behavior. They are mechanistic and repetitive and yield cycles of events that can be perceived and understood by the generalizing capacity of man. They yield daylight and darkness, tides, seasons, periodic floods and droughts that can come to be understood and timed sufficiently well to support prophecy. They can be anticipated quite accurately and behavioral adaptations can be devised to reduce the penalties these environmental forces can impose, or exploit the opportunities they provide. Man learns to return from the forest before dark, beach his canoe above the high tide mark, store food for the winter, build his house upon stilts, and take all sorts of steps that anticipate and ameliorate the effects of these repetitive cosmic forces. Where the biological environment becomes of greater concern, certain events (such as attack by marauding animals), while less predictable in timing, are sufficiently likely for anticipation to lead to taking defense precautions.

At the same time, man was discovering certain environmental relationships that were sufficiently regular and reliable to permit them to be used to amplify the power of human behavior. The earliest of these tended to be mechanical principles like the lever and the wheel or stable chemical relationships and transformations growing initially out of his familiarity with fire, water, and soil. He learned to employ them in ways that increased his physical productivity or thermodynamic efficiency. These kinds of relationships were sufficiently stable and became sufficiently well tested to permit technological prediction leading to the design of behavioral systems merging men and man-amplifying machines in the interest of improving the levels and reliability of human and social maintenance. Through the application of these organized systems of machine-like behavior, certain limited aspects of the physical environment could be modified or brought under control. Man could say, "If I organize these components of human and machine behavior in this way, I can change aspects of the environment in beneficial ways."

The two modes of prediction and planning often work together, of course. Where man's technology (i.e., his understanding of the natural forces plus the necessary amplification of human capabilities) is inadequate to control certain prophesiable events (like a flood), he may employ technological prediction and planning of a more limited sort to construct microenvironments (like a house on stilts or a boat) that will ameliorate their effects.

Differentiating the two modes of prediction and planning is a single distinction which has to do with differences in the degree of feasible environmental control. In the first case, the deterministic laws of relationship that lie behind the prophesied event and make it predictable are not of a sort that are accessible to human understanding, control, and application on a level sufficient to modify or confirm the prophecy in accordance with human purpose. For example, the cosmological relationships that govern the movements of the planets in the solar system and establish a repetitive cyclical order of events can be understood sufficiently to predict the passing of the seasons and the ebb and flow of the tides. The knowledge can be used as a basis for an anticipatory preadaptation that is favorable to human purpose— one that is defensive or amelioratory in purpose or one that exploits the prophesied event for human advantage. However, such knowledge does not permit control of the event.

In the second case, the laws of relationship are such that man can use the knowledge of them as the basis for actual modification and control of some set of environmental events. These modes of adaptation involving prediction and planning have been vastly more successful in advancing the productivity and physical welfare of man than strictly reactive modes of adaptation. The frequency and range of their application have multiplied as human technology has advanced and accumulated. Their greatest adaptive successes, however, have been in the realm of physical systems. It is with respect to these systems that the phenomenological base has been sufficiently mechanistic and repetitive to permit repeated observation and the repeated application of the same functional relationships and sequences. Where repeated observation and application did not falsify the generalizations derived

117

from observation and practice the principles came to be accepted as laws. In time, a consciously applied and organized physical science revised and strengthened these generalizations from lay experience by deliberately designing rules and tests potentially leading to the falsification of the hypotheses. It is these laws and law-like statements (together with the specification of initial conditions) that form the basis for both forms of prediction and planning.

The two kinds of prediction, of course, yield two general approaches to planning. Matched with prediction as prophecy is a mode of planning that evaluates the reasonableness of planning activities in terms of whether or not they fit in with impending changes. One plans behavior appropriate to anticipated changes in environment. Matched with technological prediction is planning viewed as a form of active predictive control of the environment.

The existence of the laws of relationship that permit these forms of prediction and planning is based upon the relationship being sufficiently deterministic and generic to support generalization and their application in the design of behavioral systems. Classical science has made its great strides by progressively discovering the domain of these deterministic relationships and applying the derived laws to the design of technological systems that amplify human behavior.

The Application of These Modes in Social Adaptation

We observed earlier that as the power of generalization gave to man a cumulating power to predict and control aspects of his physical environment, the implementation of this knowledge through behavior required the development and organization of even more complex forms of social behavior. It also permitted dramatic expansions in the scale of human population and social organization. Thus, man has found that an ever increasing part of his environment is social rather than physical. Increasingly, the problems having to do with his welfare and survival have to do with his adaptation to social environments. This implies not only the adaptation of individual behavior, but the adaptation of the behavior of more complex social systems as well. Man has been faced, then, with the need to apply anticipatory modes of prediction and planning to social as well as physical phenomena. Having been in the habit of making positive predictions and confident adaptations in the physical realm, it is only natural that he would presume to do the same in the social realm, and in the same way. His attempts to apply prediction and planning to the social process took the forms described in chapter I. In those conventional concept models of social change we can recognize both prediction as prophecy and technological prediction.

The growth models described are based upon prediction of the scale of the system, which takes the form of prediction as prophecy. In the case of the simple and allometric growth models the system is considered to be independent and the prophecy is related to endogenous forces. In the system open to resource exchange the prophecy takes into account exogenous as well as endogenous forces.

In both the allometric model and the system open to resource exchange a complex system with subsystem components is envisaged, wherein the prophesied change in scale is extended to the changes in the scale of the components on the basis of certain deterministic structural relationships or coefficients that link component scale to system scale. The model concept sometimes recognizes that a period of time elapses during which are eliminated those coadaptive responses of the subsystems that are inconsistent with the structural coefficients. Under this view, planning that takes the form of technological prediction is allowed a role. By identifying the structural coefficients, a planning agent may direct the subsystem adaptations to changes in system scale in a way calculated to bring all system components to optimum adjustment without the wasteful and time-consuming iterations that nondirected coadaptations might require.

In development models where behavioral change is allowed, prediction as prophecy and technological prediction can be seen to take the form of programmed learning. The concept of the machine system open to information also embodies both modes of prediction; here, prediction as prophecy concern itself with the diffusion rates associated with the inflows of new technology of exogenous origin, and technological prediction is seen as the intervention of a planning agent to accelerate and make more efficient the redesign of the system.

The development stage model, as we have seen, acknowledges the possibility of endogenous sources of new behavior but relates it to a

119

deterministic sequence revealed by history. In this model, prophecy takes the form of the prediction of sequences of behavioral change, rather than just the dimensions of a change in scale as in the growth models. This mode of prediction and prophecy, too, may be associated with technological prediction.

In all these conventional models of social change, the process is seen as essentially deterministic. Accordingly, prediction as prophecy is dominant in the model concept. Technological prediction, or *planning, is viewed as engineering the prophecy.*

The Limitations of the Conventional Modes of Prediction and Planning

There are short-term planning problems involving relatively simple systems where, despite their shortcomings, conventional approaches to prediction and planning can be made to work reasonably well. For example, an industrial establishment, operating within the context of a machine system open to resource transfers, might predict a change in its level of sales, then select the combination of inputs and processes that would most efficiently fulfill that level of output, then put the modified behavioral mix into operation. It may have a high frequency and degree of success with this sort of procedure.

It does so, however, because certain conditions are fulfilled. First, the sales prophecy (i.e., the change in exogenous parameter), although not certain, is more likely to prove a viable basis for planning if (1) the period over which the projection is made is sufficiently short for the historical trends to have a fairly high degree of stability,[1] and if (2) the system goals are instrumental goals that do not involve major group conflicts or require major goal adjustments.

Second, the system subject to modification must be relatively simple and the technology that defines its relationships be well known. The planning process at this stage takes the form of a technological prediction. It says, "If we change the combination of these known elements in

---

[1] Over a relatively short period, major changes in human knowledge and in environmental relationships are less likely to occur than over a longer time, and significant inertial factors are at work.

these known ways, the system will be modified in a predictable way to achieve the most efficient fulfillment of the predicted scale." It involves technological prediction but employs it as a subordinate instrument in the reactive adaptation to the prophesied scale.

Creative social learning is involved in a minimal way. No basic research or major design work is required because the technology is well established. There are no major problems of goal modification or conflict resolution, and no novel or major changes in supply channels and procedure initiated because the investments in larger stocks of human and physical capital can draw upon capital goods industries that are already organized to meet the needs of normal replacement required by social maintenance. Under such circumstances the transformation path from the initial to the terminal state raises no special problem.

The fact that something like prophecy with optimization as a planning procedure has proven useful in some applications with a minimum of conceptual complications has, we have seen, tempted more extended use. Functional social systems encounter many anomalies or malfunctional episodes. Against the background of an economist's traditional interest in optimization and the success of prophecy plus optimization as a planning concept in more restricted applications, one is inclined to see social system anomalies as resulting from the fact that the planning units are too small to take into account the external costs and benefits that accrue to higher levels of system aggregation and control. The natural conclusion is that we should attempt to apply these traditional planning concepts at higher order system levels.

A number of examples of this attempt are to be found in the economics and planning literature. From among them let us choose for examination the emerging practice of urban planning.

There is a tendency in some quarters to look upon urban planning as the specification of an optimum network design for a complex urban system and, consequently, to regard it somewhat in the light of an engineering design problem. The city of the future has to be put in place, and the way in which the activities are to be bundled and the transfer channels planned is the problem to be solved. This is a logical extension of the efficiency optimum of static state equilibrium systems common to the tradition of economics. It is the model of the machine system open to resource exchange.[2] First one projects the parameters that establish the future levels of the activity components, then designs the urban system that most efficiently organizes these activities into an appropriate network.

PRACTICAL OPERATING DIFFICULTIES

The current literature occupied with the explanation and practice of this mode of prediction and planning is already reflecting some of the obstacles to this approach. It is commencing to make plain that for large systems the computational complexity of this kind of design problem is staggering and that even our present short-cut methods for systematically searching out optima cannot cope with problems of this dimension. (See Werner, 1968; Harris, 1967; and Lowry, 1967.) Equally staggering would be the task of carrying the research of real process-scale economies and bundling options to the point that all of the technical coefficients would be available that are necessary for the solution of an optimum network design problem. One cannot assert with confidence that the time will never come when research carried out in this frame can yield useful results, but the prospects for near-term success are sufficiently dim to discourage the application of the substantial effort that would be required to move ahead with this prob-

---

[2] In traditional location theory a restricted network optimum is implicit in the solution of the minimum cost problem, although it does not make explicit the problem of designing networks. Central place literature only goes so far as to identify deductively the advantages of a generalized hub and spoke network for multiple purpose network circuits. The problem is not characteristically formed as a general optimization problem in the early literature. Some later attempts have been made to link the central place ideas with the optimization solution as viewed by the general equilibrium theory in economics. (For example, see Stephens, 1968.) However, this formulation was addressed to the distribution of land between competing functions through the operation of rent gradients. In these cases the problem of optimum network design has been only partially considered although the orientation is there. A more direct interest is beginning to emerge in the recent work— particularly that of some of the analytical geographers. Notable among these are Beckmann, Friedrich, Garrison, and Werner. See Werner (1968) for an excellent example and references to the work of others. Harris (1967), on urban planning, has also displayed an interest in this problem.

lem. In addition, it is apparent that even if an efficiency optimum could be achieved it may not serve the adaptability that is essential for the system to function well as a learning system. This is especially apt to be true when the attention is focused largely upon the networks of transformations and transfers characteristic of the physical processes related to social maintenance.

Lowry (1967, pp. 197–98) has offered a succinct statement of these problems in commenting upon the search for an optimum urban network design: "Let me ... suggest that the search for an *optimum optimorum* may itself be a misallocation of resources." He offers two reasons. First, "... I submit that the designer/planner's failure to blueprint the best of all possible worlds is at least partly offset by adaptive behavior of people within the orbit of the system which may have been suboptimally designed. ... The other reason for my lack of concern ... is that the external constraints visible to the designer will be changed by the time the design is implemented. ... These considerations lead me to the conclusion that a wise designer should be prepared to sacrifice a good deal of visible efficiency to future flexibility. ... In the design of large systems I would argue that the scientific analysis of system mechanics will enable us to avoid large mistakes even though it will not enable us to locate *optimum optimorum*."

Interrelated social networks are not put in place intact; they evolve. They develop by adding new channels, reorientating old channels, modifying channel capacities, and superimposing new network segments upon old. This is a consequence of the fact that social systems behave as learning systems. A capacity for creative learning is essential for the transformation of social systems.

In short, the application of these conventional modes of prediction and planning encounters a number of practical limitations. Economists and operations research specialists who like to operate on the basis of the conceptual mode characterized as a system open to resource exchange have experienced a number of operating difficulties. Some of the difficulties are identified as follows: (1) The structural coefficients that define the relationship of a system's components may not define a single unique efficiency optimum. (2) Information sufficient to permit such a solution is commonly not available—especially for large-scale complex systems like urban and national systems. (3) Even without

123

information problems, computational methods and tools do not have the capacity for solving such complex multivariate problems.

There is sometimes a tendency to assume that persistent effort to improve information sources and computational capacity will one day permit better performance under the control of this conventional mode of prediction and planning. However, the most serious operating difficulties stem from a different source. The available technology or behavioral options are subject to change. The temptation here is to switch to a more general model characterized earlier as the system open to information exchange. This model is based upon an image of an exogenous society that has already developed techniques and behavioral designs that contain the potential for improved system performance if adapted. The realization of this improved performance requires a diffusion of the technology and its embodiment in artifact machines and human behavior in all those social subsystems for the behavior of which the technique is relevant.

This approach to prediction and planning makes more clear the practical limitations of the conventional modes.

First, the operating subsystems are, in the end, human beings acting as individuals and groups. They have the option of refusing to be reprogrammed with new techniques and frequently exercise that option.

Second, in such a situation there is less likely to be a unique solution that defines a new optimum terminal state for the system. Because the new techniques cannot be introduced to all system components simultaneously (a practical impossibility in complex systems), one confronts the reality that diffusion of the new techniques defines a time-space path as it spreads through the system. There are a number of different pathways that a set of new ideas may take depending upon when, where, and how they are introduced to the system. Each may exhibit a different rate of diffusion and a different terminal solution. If the problem presented by the planning solution of a complex system with fixed technology can be too large for solution, then the complexity and attendant difficulties are vastly multiplied when we try to plan for a terminal solution implied in a new, exogenously available technology.

Third, the key limitation is the fact that all social systems are human systems. This implies that they contain an endogenous capacity for learning and behavioral change. Social system development is not just an active or reactive reprogramming of the behavioral components of the system. It also emerges out of internal problem solving and creative play. Indeed, even the successful adoption of new behavioral ideas of exogenous origin frequently requires a creative learning modification. Adoption is linked with adaptation.

The reproduction of prototype behavior in human beings through training is always subject to the emergence of unsuspected idea mutations generated by the creative capacities of the human beings involved. Further, a behavioral idea developed and successfully tested in one system setting may not function quite as anticipated in another—even though they be similar in many respects. This can give rise to functional anomalies that require new idea development and implementation to resolve. The new or modified design idea that is untested in system behavior does not certainly or deterministically fulfill the system objective. It becomes a hypothesis that has to be tested through its embodiment in system behavior. It may make a verified contribution to the system goals or it may fail, to be replaced by still another novel behavioral idea or a refinement of the discredited one. Creative adaptation and open-ended social learning inevitably invade the process.

It is precisely these creative sources of behavioral change that are not taken into account in the conventional concepts of prediction and planning.

LIMITATIONS BASED UPON THE DIFFERENCES IN THE NATURE OF SCIENTIFIC LAWS WHEN APPLIED TO EVOLUTIONARY PHENOMENA

There is a second way that we can consider the limitations of the conventional modes of prediction and planning. They try, implicitly or explicitly, to apply to the analysis and management of social processes the concept of the scientific law in a way that is inappropriate to the nature of the process.

Prediction and planning under the control of the conventional concept discussed assume that the analyst-planner is dealing with an essentially deterministic process so that the change in the behavior of the system is defined by certain laws of relationship and certain laws of dynamic sequence or evolution. But the scientific study of the social process does not yield laws of this character in the same ways as the experimental study of physical systems does in classical science. Let us consider this matter, beginning with the idea of a law of sequence.

The concern with prediction as prophecy in the conventional sense has given rise to an almost compulsive interest in ascertaining the "Law of Evolution" or "Law of History" or a "Natural Law of Succession." An extraordinary amount of attention has been devoted to the identification and proclamation of such a law. Man desperately wants to anticipate the course of social transformation in the hope of thus determining basic criteria for governing human behavior and social adaptation. But one thing is abundantly clear: *The scientific study of biological evolution does not support the conclusion that there is a Law of Evolution* (for either the biosystem or society as a whole) capable of providing us with an instrument for foretelling the future. In this sense, long-run prophecy is scientifically impossible. There is no such thing as a Natural Law of Succession.

The reasons are fairly simple, and Popper (1964, pp. 108–9) has elucidated them with admirable clarity. "The evolution of life on earth, or of human society, is a unique historical process. Such a process, we may assume, proceeds in accordance with all kinds of causal laws.... Its description, however, is not a law but only a singular historical statement. Universal laws make assertions concerning some unvarying order ... concerning all processes of a certain kind.... But we cannot hope to test a universal hypothesis nor find a natural law acceptable to science if we are ever confined to the observation of one unique process."[3]

Evolutionists and historicists may counter such a conclusion by maintaining that there is an overall trend in evolution that operates like an inertial factor and can serve as the basis for long-run prophecy. To again quote Popper (1964): "The answer is: trends exist, or more precisely, the assumption of trend is often a useful statistical device.

---

[3] The same point about the absence of a warrant for speaking of a "law of evolution" or its "overall direction" is made by other philosopher-scientists as well. (See Goudge, 1961, pp. 125–26, 168–79.)

But trends are not laws. . . . a statement asserting the existence of a trend at a certain time and place would be a singular historical statement and not a universal law. . . . while we may base positive scientific predictions upon laws, we cannot (as every cautious statistician knows) base them merely upon the existence of trends" (p. 115). "Explained trends do exist [i.e., when they become components of conditional law-like statements], but their persistence depends upon the persistence of certain specific initial conditions. [The evolutionists] *overlook the dependence of trends upon initial conditions.* . . . This, we may say, is [their] central mistake. [*Their*] *'laws of development' turn out to be absolute trends*; trends which, like laws, do not depend on initial conditions and which carry us irresistibly in a certain direction in the future. They are the basis for unconditional *prophecies,* as opposed to conditional scientific *predictions*" (p. 128). The persistence of the initial conditions cannot be assured nor the nature of their transformation prophesied, because "We cannot predict, by rational or scientific methods, the future growth of our scientific knowledge" (p. vi). We cannot prophesy about the social process beyond the horizon where major new increments to our knowledge may arise.

At the same time, we are not warranted in concluding that evolutionary explanations are not scientific or that we are completely unable to produce any useful predictive statements applying to the behavioral transformations of populations (either social or biological) acting as open learning systems.

The reason for the first proposition is the fact that the test of a scientific statement[4] is not that it must have positive predictive value, as many people suppose, but that it must be susceptible to falsification by evidence. As Goudge (1961, p. 65–79) makes plain, there are integrating and narrative explanations of historical phenomena that constitute legitimate scientific statements because they are empirically testable—even though they do not support positive predictions. They enter into evolutionary theory at the points where singular or unique

---

[4] The test of a scientific statement is what the science philosophers refer to as the "problem of demarcation"—the test that distinguishes scientific statements from metaphysical ones.

events of major importance for the history of life need to be explained and where direct evidence linking subsequent and antecedent events is incomplete. They are illustrated by the explanation in chapter II (pp. 54–55) of how the vertebrates first came to occupy dry land. While such explanations do not support positive prediction, a negative prediction is implied in their form. Such narrative statements imply that future research *will not* turn up evidence inconsistent with the explanation. They are scientific because they are susceptible to falsification and such statements play an indispensable role in evolutionary theory. It seems plain that present and future research employing the materials of social history in the interest of understanding social evolution will have many occasions to legitimately apply such narrative explanations.

More important with respect to our immediate problem is the fact that there is still a role for prophecy in evolutionary theory—albeit a circumscribed one. The fact that biological and social evolution "in the large" is a unique historical process does not mean that it has been impossible for evolutionary theory to develop any law-like statements. At the level of phylogenesis in the biological realm (as distinct from a concern with the evolution of the biosystem as a whole) the evolutionary process has taken a number of distinct lines. To the extent that each lineage has been subject, in its transformation, to similar sequences or preconditions, it has been possible to develop a number of law-like statements.[5] One such statement, for example, asserts that genetic drift occurs in small populations. Such a statement is law-like because it generalizes about a number of instances. Yet such statements cannot claim the status of universal laws, nor do they support firm positive predictions. (1) They are not subject to as many opportunities for falsification; and (2) the opportunities for testing that do present themselves are not subject to perfect control or matching of initial conditions after the fashion of laboratory testing, so the results are not completely unambiguous. However, they can support a form of prediction that is useful and justifiable. Goudge calls these "hypothetical probability predictions."

He illustrates this point in the following fashion (pp. 126–27):

[5] Goudge (1961, p. 122) offers a sample of such statements.

"Suppose a small isolated population ... is discovered. By virtue of a law-like statement which asserts that genetic drift tends to occur in such populations, we can predict that (the population) will probably be poorly adapted and that its chances of avoiding extinction for very long are small. This prediction is far less securely grounded than any astronomical forecast. Yet it is certainly more than a guess. It has some foundation in the available evidence. ... our prediction depends upon the acceptance of a law-like statement. ... Yet what we forecast is conditioned by a large number of factors (e.g., the future state of the population's environment, its mutation rate, etc.), and because of our limited knowledge we cannot provide anything like a full inventory of them. Hence our prediction must of necessity be an approximation having a substantial margin of error. So it is with all prediction that modern selectionist theory makes possible." [And] "only a limited number of evolutionary phenomena ... are amenable to predictive treatment."

Thus, modern selectionist theory has generated a number of law-like statements pertaining to the processes of adaptation common to all species subject to the learning process we know as phylogenesis. These statements indicate that under different conditions a species may generalize or broaden its adaptive range, specialize or narrow its adaptive range, or fail to adapt sufficiently and become extinct. The illustration above taken from Goudge pertains to the conditions under which the latter prospects become likely. The reader will recall from chapter II (p. 54) that Simpson has similarly identified the preconditions for adaptive generalization in phylogenesis.

Again, such law-like statements are not sufficiently developed and tested to form a firm prognosis for a species that has exhibited all of the preconditions for adaptive generalization and has, as a consequence, broadened its behavioral capacities. We can state with considerable assurance, however, that such a result was a great deal more likely than in the case of another species for which these preconditions were not fulfilled. Furthermore, in negative form the predictive statement becomes stronger. We might say that "The species that does not fulfill certain conditions for adaptive generalization will not increase its behavioral adaptability." In short, the incompleteness of our knowledge about the behavior of open systems or learning systems may enable us to hypothesize some of conditions *necessary* for the transformation without being able to include in the hypothesis all of the conditions *sufficient* for that result.

This may serve to recall to the reader the nature of the development model implicit in biological phylogenesis and articulated at the end of chapter II. It was characterized as a stochastic learning model. The development path of an emergent species is based upon the interaction of two sets of conditional probabilities—one related to the development options conditioned by the history of the species, and the other related to the environmental variation the species might encounter. The fact that Goudge (1961) characterizes the generalizations in biology as "law-like" and prediction as "hypothetical probability predictions" is a reflection of the stochastic nature of the process.

There is, of course, no conscious planning in the phylogenetic process, but man has learned to base technological prediction upon these conditional probabilities where they can be given some empirical dimension. In these cases planning becomes a form of betting upon conditional probabilities. This is what the plant or animal scientist does in carrying out controlled breeding.

Just as the life sciences have developed some "law-like" statements of sequence, they have also developed some law-like statements of structure. There are, for example, certain characteristics of organization that different ecosystems tend to have in common. An emergent ecosystem would be likely to possess some of the same characteristics. In a similar way man may use such knowledge as a basis for placing planning bets. As knowledge of ecosystems advances, this knowledge may be applied to efforts to control some elements of the ecosystem so that the coadaptive responses of component species may not lead to the extinction of desirable or scientifically important species.

It is obvious that generalizations that relate the behavior of systems in the social process can (like those related to the biological process) claim to be no more than "law-like" and predictions based upon them can be no more than probability predictions. If one assumes that a similar process of study and generalization is available to social science, this suggests a modification of the procedures of study in social science away from those guided by the conventional model concepts we have discussed. A movement in this direction is already discernible.

In the development of law-like statements of sequence, two forms may be observed. The forerunner of the first is seen in the popularity of stage theories of development. Where stages are identified in history and interpreted to be unilinear and deterministic, they amount to an attempt to identify a "law of history" and are subject to the same objections and limitations already noted. However, it may still be possible to study economic and social history with a view to identifying sequences that seem to be common to different kinds of historical initial conditions and, perhaps, even go so far as to derive from this historical record some idea of empirical probabilities. Gerschenkron (1962) is an example of a movement in this direction.

There is a second source of probability statements about sequence that is unique to the social process. Because of man's active, anticipatory, analytical capability, he can to a limited degree anticipate history. He can deduce something about the future behavioral design implied by already existing knowledge. The recent developing interest in technological forecasting is an example of such an approach.

In short, law-like probability statements of sequence can be generated by both inductive and deductive procedures. However, work in the social sciences directed explicitly to producing these kinds of generalization and under the control of the more general conceptual model is, to date, relatively limited in scope and recent in origin.

Similarly, social science may seek to develop law-like statements of structure. In this connection much additional work needs to be undertaken to support greater understanding of the patterns that are available and to the circumstances under which the different network patterns are most likely to appear, as well as to organize the transformations and transfers that make up social activity.

There is no doubt that this evolutionary metaphor and the model concept it provides can serve to organize and control some very useful social science research. It is also plain that its implications for research have not been adequately exploited. However, when we have understood the unique characteristics of the social process developed in the last chapter, we must recognize that the statements generated by such research cannot be law-like even in the sense or to the degree that characterizes the life sciences.

There are two reasons why this is so. The first stems from the fact that a social system is less precisely and unambiguously bounded, than a biological organism, species, or ecosystem. Thus, research that tries to generalize about the organization and development of social systems of similar kind is faced with much more difficult questions of identification and taxonomy than in the life sciences. Furthermore, the boundary conditions that define a social system are subject to substantial modifications over much shorter periods of time. This has the practical effect of making generalization less secure in the social sciences than the life sciences.

More important for our purpose is the fact that probability predictions based on law-like statements in the social sciences may be either self-fulfilling or self-defeating. Individuals and social groups make plans and base actions upon the probability prediction and the action serves directly or indirectly to modify the probabilities that define it. This is much more than the progressive transformation of a random probability matrix, as would characterize the development model as a stochastic learning process (chapter II). It is a nonrandom and sometimes purposively directed change in the probabilities.

If a probability prediction concerning future states of a system is viewed as a conditional prophecy by system components for whom total system control is out of the question, their defensive adaptations, no one of which would have any influence, might collectively modify the probabilities in a nonrandom fashion. The economist, for example, is very familiar with the way in which a widely disseminated probability prediction of inflation can lead to component behaviors that confirm the prediction or increase the probability.

If the probability prediction is embodied in a technological prediction by an agent capable of acting directly upon the system, then the prediction itself may be made the direct object of purposive modification. If the prediction serves to anticipate a problem not yet manifest, it may prompt a planning action addressed to forestalling the problem and consequently will modify the prediction.

There is another way of expressing the inadequacy of the metaphor of the stochastic learning process. The reader will recall the Slobodkin interpretation of biological evolution in terms of game theory (chapter II, p. 70). Game theory in its present stage of development is primarily absorbed with the determination of the most advantageous strategy in play, given the objective of the game and the nature of the rules of play. But this is an extremely narrow interpretation of the vital nature of the social process. The most critical problems in social change are not those associated with the selection of an optimum game strategy; they are more often those connected with the necessity for or desirability of changing the rules of the game. We are at critical points faced with the problem of choosing the most advantageous game rather than the most advantageous play under the control of a given set of rules.

Once again we find ourselves in a new realm not accommodated by the conventional approaches to prediction and planning or, indeed, by the more general stochastic learning model. We come face to face with the purposive creative process of social learning. We are faced with the necessity to identify the nature of prediction and planning under the control of the concept of social learning. Indeed, all of the discussion to this point has been undertaken with the intent of preparing us for a more detailed exposition of the nature of this process.

Prediction and Planning as Evolutionary Experimentation

Social learning involves more than the engineering of prophecy or the playing out of an optimum strategy in a stochastic game with fixed rules and objectives. Social learning is much involved with changing the "name of the game." At root, the learning process is manifest as an iterative exploratory series of experiments in social action. Although uncertainty or a stochastic dimension remains an important element in the process, the conditional probabilities or elements of uncertainty are made an object of deliberate, purposive modification. There is a directed normative component to this process that is supplied by human goals.

We characterize this social learning process as evolutionary experimentation—a phrase first encountered in a work of Sackman's (1967). The concept is similar to one which Popper (1964) calls "piecemeal problem solving," although Popper does not fully articulate the nature of the concept in its social context and, as will be apparent, does not treat the full scope of the process.

Prediction takes the form of a developmental hypothesis, which may be expressed as follows: "We hypothecate that if we undertake a certain course of action, the performance of the social system will be modified in such a way as to improve its efficiency in satisfying the goals or objectives of the system." This is not known to be true on the basis of established deterministic laws. It is not an exercise in simple engineering design. It is an experimental design. Planning takes the form of conducting an experiment by embodying the new modes of behavior in the performance of the system. It can be viewed as testing the developmental hypothesis in action. If the developmental hypothesis is not falsified by the results of the experiment (i.e., if the performance of the system with reference to its goals is improved), the novel mode of behavior will be reinforced and persist. If the results call the hypothesis into question, a new or modified one will take its place and a new experiment in social action carried out.

It is not implied that social learning is always the product of a consciously designed and carried out social action experiment. During much of social history to date this has not been so. Often a proposed change in social action is not recognized or conceived as a developmental hypothesis. Man has a strong tendency for self-deception concerning the efficiency of his social nostrums. Even where free from self-deception, those who are engaged in initiating a change in behavior often feel it necessary to represent the change proposal as a certified cure for social ills or a certified instrument of social gain. Man often acts as if the validity of a developmental hypothesis is preestablished through faith.

As a consequence, the implementation of social action is rarely viewed as a test. Indeed, it is often carried out in such a way as to obscure the results of the test and render difficult, if not impossible, dispassionate evaluation. Ineffective social action may persist for some time because it is not recognized as such.

Even so, conscious design or no, the practice of social innovation throughout history has refined a set of practices that survive and are generally appropriately related to social goals. Though novel social action is not designed as a test of developmental hypotheses, ex post facto hypothesis falsification frequently comes to light and clearly ineffective modes tend to lose out. Thus practiced, there is a great deal of inefficiency in the process—a matter which will concern us shortly when we inquire whether there is some way to make the learning power more efficient. Nonetheless, evolutionary experimentation is the fundamental character of the process of social learning.

Evolutionary experimentation involves hypothesis testing of a different order from that associated with classical experimental science. The latter evolved with the study of physical systems where the relationship between system variables is commonly highly deterministic, and where the concern is more often addressed to the effect upon the behavior of a system component, or component system, of a change in an exogenous parameter—usually under highly controlled conditions. Social problem solving, on the other hand, is less concerned with the effect of a change in a system parameter upon a component than it is with the effect of a planned change in a system component upon total system behavior.

In physical systems classical science is concerned with the consistency and universality of the relationship. The same hypothetical relationship is tested repeatedly under nearly identical conditions; the more often it survives falsification, the more soundly the principle is deemed to be established. Where tested consistency is high, the relationship is identified as a law which may be employed in limited prophecy and in holistic engineering design.

The social experiment that takes the form of the implemented plan is not subject to this kind of testing and does not seek to establish universal principles. It cannot because it is phenomenologically unique. At any one time any functional social system displays a unique behavioral state formed under the influence of a unique set of environmental relationships and a unique history. The same planning experiment cannot be performed upon two identical systems. Furthermore, the object of the planning experiment is to change the performance of the system. One cannot return the system to its preexperimental state and retest the same experiment or comparison-test an alternative planning hypothesis. To achieve this, social problem solving engages in experimental tinkering with the system by subjecting it to a series of behavioral changes each of which is evaluated in terms of whether or not it improves the system's overall performance with respect to some

135

desired goal or criterion of behavior. The fact that this mode of experiment does not yield universal laws makes it no less scientific if it is consciously and objectively carried out and subjected to the test of performance.[6]

In short, prediction as hypothesis and planning as experimental design are the fundamental core of social learning. It is this rudimentary and fundamental form of prediction and planning out of which social knowledge and environmental control have evolved. The goals of this process are not predetermined by history, but are provided by man as a fundamental component of the predictive hypothesis; history only helps define the constrained realm of the possible and the valuable. This remains the root method of social progress. Through this process man is acting to invent history.

NORMAL PROBLEM SOLVING VERSUS THE PARADIGM SHIFT

Our understanding of this process is enhanced if we recognize that it exhibits both a normal and an extraordinary mode—a concept that is ably explored by Kuhn in *The Structure of Scientific Revolutions* (1964). Here Kuhn, a physical scientist who devotes his attention to the history and philosophy of science, provides a fascinating analysis of the nature and causes of revolutions in basic scientific concepts.

The great bulk of scientific research in all fields is what Kuhn characterizes as "normal science." This is the orderly cumulative work of science that takes place under the unifying control and direction of a "paradigm." The paradigm is a metaphor (not considered as a part of language but as a framework of thought) or unifying theory

---

[6] The above may oversimplify the process. It contrasts evolutionary experimentation with a historical and still widely held view of the nature of the scientific method and classical laboratory experimentation. It does not argue against the unity of method claimed by Popper (1934, 1964). The physical sciences are coming to recognize that even in the physical world the results of their experiments are not independent of the presence of the observer or experimenter (the Heisenberg principle). And Popper's view of the "logic of scientific discovery" can be interpreted as an evolutionary epistemology. (See Campbell, forthcoming.) However, the present volume, that of Sackman's, and the writings of Kuhn (1964) and Schon (1963) carry the description of evolutionary experimentation along a somewhat different path from that of Popper.

that has emerged out of earlier scientific practice and supplies the foundation for its further practice. Men whose work is carried on under the influence of a common paradigm are committed to the same standards and rules for scientific practice. Paradigms become established because they are successful in solving problems that a group of scientists recognize as important. Each paradigm, however, leaves room for a great deal of mop-up work essential for realizing its promise. That promise is fulfilled by "extending the knowledge of those facts that the paradigm displays as particularly revealing, by increasing the extent of the match between those facts and the paradigm's predictions, and by further articulating the paradigm itself" (Kuhn, 1964, p. 24).

This work of normal science takes on the nature of "puzzle solving." The results gained "are significant because they add to the scope and precision with which the paradigm can be applied." The general acceptance of the paradigm and its previous successes give a high assurance that the remaining puzzles have a solution and spur on the scientific endeavor within the domain. Each of the scientific disciplines and their distinctive subspecialities is carried on under the direction of a paradigm. They have a hierarchical character so that one paradigm may have a controlling influence over the activities in several subdisciplines, each functioning under the control of subsidiary paradigms.[7]

One of the striking things about the conduct of normal science is that it does not *aim* at producing novelties and when it is successful in its own terms it finds none. It seeks the progressive testing and refinement of a paradigm until its correspondence with nature is perfected. Yet there are times when this very process serves to generate novelty. This is best described in Kuhn's words (pp. 52–53).

New and unsuspected phenomena ... are repeatedly uncovered by scientific research, and radical new theories have again and again been invented by scientists. ... If this characteristic of science is to be reconciled with what has already been said, then research under a paradigm must be a particularly effective way of introducing paradigm change. That is what fundamental novelties of fact and theory do. Produced inadvertently by a game played under one set of rules, their assimilation requires the elaboration of another set. ...

[7] The basic reference (Kuhn, 1964) is rich in examples of normal science at work to which the reader is advised to turn for further illustration.

Discovery commences with the awareness of anomaly, i.e., with recognition that nature has somehow violated the paradigm-induced expectations that govern normal science. It then continues with more or less extended exploration of the area of the anomaly. And it closes only when the paradigm theory has been adjusted so that the anomalous becomes the expected. Assimilating a new sort of fact demands a more than additive adjustment of theory, and until the new adjustment is completed—until the scientist has learned to see nature in a different way—the new fact is not quite a scientific fact at all.

Thus, in the life of every science the practice of normal science at times turns into extraordinary science. Then, some scientists turn their attention to formulating competing theories and paradigms. Those that the community comes to agree upon succeed the older paradigms and there follows a consequent change in the rules and standards of investigation that govern the practice of normal science in that discipline. The anomalies of normal science create crises that lead to scientific reevaluations. One example is the replacement of Aristotelian physics by Newtonian physics and its replacement, in turn, by quantum mechanics.

Each paradigm shift brings a newly emerging world view to the scientific practice it controls. Following the shift is a relatively prolonged period during which normal science is largely devoted to filling out, testing, and extending the understanding implicit in the paradigm.

Kuhn is not alone in articulating this concept of the paradigm shift. Writing independently and at about the same time, Schon (1963) worked out a very similar representation based on his study of the process of technological invention. The terminology is different but the concepts are nearly identical. Where Kuhn speaks of a "paradigm," Schon speaks of a "metaphor." Where Kuhn speaks of a "paradigm shift," Schon speaks of a "displacement of concepts."

Schon takes as his problem the explanation of how new concepts arise, pointing out that "in working toward the new, we have nothing to use but the old" (p. 22). The established metaphor is taken "as a *programme* for exploring the new situation" (p. 59). In applying an established metaphor to the interpretation of a new instance the tendency is to apply the metaphor literally to the new situation—to see the new strictly in terms of the old. This would appear to be similar to the puzzle solving Kuhn characterizes as "normal science." But frequently the application of the established metaphor to the new situation changes the way in which the metaphor itself is perceived: ". . . the old theory becomes new as it changes to meet the demands placed upon it as a projective model for restructuring the new situation" (p. 64).

Schon goes further than Kuhn in trying to explain the process of concept displacement or paradigm shift. For example, he points out that

Metaphors that come to mind and are selected result from the interaction of what is given or imposed by the culture with the demands of the situations confronting us. Theories are chosen from among the culture's store, on the basis of their ability to meet these demands—giving rise to metaphors which will explain these troubling situations or allow them to be changed. (p. 68)

Metaphors also come to be selected out of the fact that new situations are never wholly without conceptual structure. Any situation for which we attempt to develop new theories already has theory-structure of a sort. So there is a tendency to select metaphors which are already implicit in the language of the theory, or which mesh in certain ways with the metaphors underlying the theory. (p. 71)

Both of these treatments are rich sources and well worth the careful attention of the reader.[8]

These interpretations of science history and the development of technology add to our understanding in several ways. First, we can see that this same distinction applies to the process of evolutionary experimentation in a social system context. Kuhn was primarily concerned with the evolution of concepts in the domain of classical science devoted to the analysis of physical and biological systems, and made no effort to interpret the process in a social context. But the parallel is inescapable.

[8] There are other parallels in Schon's treatment that are suggestive. For example, in reviewing the conventional theories of novelty or new concept formation (pp. 12–21) he identifies (1) "theories of mystery"—that are like the orthogenetic theories in biology and their social science counterpart identified later in the chapter, (2) "theories of reduction"—similar to deterministic programmed learning, and (3) "theories of scientific method"—where the "logic of discovery turns out to mean the logic of justification of new hypotheses in science" and where the issue of new concept formation is not really joined.

A social system or man-machine system acting as a behavioral system is characterized by the organization of component behaviors in the pursuit of a set of objectives. If this were not so it would have no entity as a system. Since system objectives are often not adequately fulfilled by system behavior, the resulting anomalies give rise to the application of evolutionary experimentation. New ideas are developed and implemented that modify the behavior of the system in a way, if successful, that narrows the gap between objectives and performance. This is normal problem solving which takes place under the guidance and control of the accepted notion of the system entity and objectives.

In short, the recognized functional boundaries of a social system (i.e., the accepted image of its entity) and the functional objectives of the system constitute a "paradigm" or "metaphor" under the control of which the component activities are carried out. Normally the problems or anomalies that are revealed by the operation of the system are solved by experimenting with developmental hypotheses consistent with the general notion of social system entity and objectives.

But sometimes such behavioral innovations fail to improve the performance under the terms of the paradigm or system objectives. When this circumstance arises, evolutionary experimentation may take one of two paths. The usual reaction is to reject the innovation. The experiment is judged to have failed. A new or modified developmental hypothesis arises and a new test is performed under the control of the accepted concept of the social entity and social system goals. The behavioral changes that are initiated are characteristically changes in the performance of the social system components that do not modify the basic entity of the system and are designed to improve the system performance relative to accepted goals. Social reorganization and behavioral change are primarily endogenous. The only goals that are modified are those goals or criteria that are instrumental to the fulfillment of primary system goals.

But sometimes the failure of a developmental hypothesis to test out, or a series of such failures, may, itself, seem anomalous. There may be grounds for believing that the problem to be solved or the opportunity to be exploited is not amenable to successful treatment within the constraints of the existing paradigm. The functional entity of the system, the system objectives, or both, may not support the kind of behavioral change essential. This may require the creation of a new or modified social entity or a revision of the major social system goals. The result is a paradigm shift in social organization and goals. Once this shift is negotiated, evolutionary experimentation may once again be absorbed for a time with the further refinement of system performance, given the new goals and the refined entity concept.

An example may help. In earlier days, the production of ferrous metals in this country was carried on by a number of identifiable social systems or establishments each characterized by a different set of objectives. Some were engaged in the production of pig iron, some converted pig to steel, some produced basic shapes and plates, etc. Through operating experience, inefficiencies in processes fulfilling these objectives became apparent. Some could be resolved within the context of the framework of existing system boundaries and objectives (e.g., the substitution of the Bessemer converter for the old puddling furnace). However, some could not be resolved without the revision of system concepts and objectives.

Substantial inefficiences, for example, were associated with the heat losses that occurred between the production of pig iron in one stage, the production of steel in another, and the production of sheet and structural shapes in another; but the problem could not be resolved within the framework of the existing establishments. Indeed, the very fact that the diseconomy or anomaly was external to the operating social systems caused it to be overlooked for some time. Attempts to cope with the problem within the framework of the existing system objectives were bound to fail. But, if one modified the system concept to accommodate a series of integrated processes within one establishment with a more inclusive or higher order set of objectives, evolutionary experimentation could and did resolve these anomalies.

It is important to note that the nature and scope of the required reorganization are revealed through the operation of the normal problem-solving process. But during these periods of social reorganization normal problem solving takes second place to extraordinary problem solving, just as Kuhn shows that normal science is temporarily displaced by extraordinary science. During these transitions the problem-solving effort is wrenched out of its customary system orientation and control and comes to consider the adequacy of the customary system

141

boundaries and objectives. Nevertheless, in such a case social reorganization and novel system behavior bear a distinct relationship to
their component and antecedent systems. *It is a logical outgrowth of
the normal practice of evolutionary experimentation.*

The economist has long been conscious of the importance of "external economies." Taking advantage of these economies frequently
involves bringing several independent control systems under the hier-
archically, superordinate control of a higher-level system. The development of integrated steelmaking is a case in point. But sometimes the
paradigm shift or the emergence of a novel social system entity takes
the form of spinning off a new, quasi-independent system. For example, a manufacturing enterprise producing assembled electronic
systems like radios may face problems in efficient set design and performance that are resolved by the development of solid state devices
like transistors. But the production of these devices may require fundamental changes in the objectives and organization of the enterprise. In
this case, however, the reorganization problem may be sufficiently
great and the requirements for internal system control sufficiently weak
to bring about a different result. Social reorganization may take the
form of spinning off the production of solid state devices in a quasi-
independent social subsystem.[9]

[9] The discussion to this point may suggest to the reader that normal evolutionary
experimentation yields changes that are essentially endogenous to the problem-solving system, while evolutionary experimentation leading to paradigm shifts produces
largely exogenous changes. This is not the case. In fact, both normal and extraordinary problem solving may be directed to both endogenous and exogenous environments.

Normal problem solving under the control of an accepted concept of system
entity and system purpose may take the form of a purely endogenous modification
of behavior. The Bureau of the Census or an insurance company, for example,
might apply the computer to a more efficient performance of their conventional
information handling functions. This would bring about changes in subsystem behavior and organization without altering social system entity or purpose. Alternatively, normal problem solving under the control of an accepted system entity and
purpose might act upon the exogenous environment. A business firm, for example,
might decide that it is failing to fulfill its system objectives as well as it would like
because of the competition of an external enterprise or the interference of some
government conduct rule. Without changing its system entity or purpose, it might
modify its behavior in a way designed to act upon these exogenous environmental

In sum, evolutionary experimentation has two modes. Normally it is a process of social system refinement under the control of accepted system goals. This normal practice, however, occasionally reveals anomalies or new knowledge that leads to a paradigm shift or social reorganization. When this occurs the concepts of system entity and system goals are, themselves, revised. The developmental hypothesis takes the form of hypothecating that a different form of game from the one currently played might prove superior.

THE ACCEPTABLE ROLE OF CONVENTIONAL APPROACHES TO PREDICTION AND PLANNING

This, then, is the nature of prediction and planning viewed as evolutionary experimentation and under the control of the social learning metaphor. It is more general and less constrained to special situations than the modes associated with the conventional models of growth and development. At the same time, it does not replace or render obsolete these antecedent concepts. They still have an acceptable role to play. One contribution that the social learning concept can make is to clarify the nature of that role.

---

systems so as to influence changes in their behavior favorable to the better realization of endogenous system goals. Normal problem solving comes to embody game strategy relationships with external systems.

Similarly, problem solving leading to a paradigm shift can operate upon either exogenous or endogenous environments. We have already observed how a paradigm shift may lead to the emergence of higher-order levels of social system control so that external economies come to be internalized. A shift in the very concept of social system entity takes place that results in the generation of either a new quasi-independent social system or a new hierarchical level of social organization. In either case some element of the exogenous world of events is brought under system control.

It is also possible that a social system undergoing a paradigm shift might generate behavioral change that is primarily endogenous in nature. For example, the electronics firm cited above might decide to produce the transistor as a part of its internal function. The effect would be a major shift in the defined purpose or goal of the system, but one that is carried out by internal changes that do not generate a discontinuity in the total system entity. It might lead to the generation of entirely novel subsystem components. These might alter the conventional boundaries of the enterprise and change its image somewhat, but the social system would be altered by endogenous behavioral changes that progressively alter its functional image.

We are now accustomed to the posture taken by prediction and planning under the control of a more deterministic paradigm. Here, prediction is conceived as a basis for total system optimization. It exhibits the intention to develop a positive prophecy of the key dimensions of the social system—i.e., it seeks to foretell the principal parameters that can serve as scalars in designing a terminal-state optimum.

The realm within which this approach to social change can be realistically applied is limited not by the incomplete state of scientific knowledge, as many seem to think, but because the process of social change implied by this concept finds realistic application only under special conditions. It is conceived, that is, as being a process leading to a predetermined terminal state that leaves only efficiency choices to the planning process.

However, we can now see that these less general and more deterministic modes of prediction and planning may be governed in their application by the practice of evolutionary experimentation. They are not independent or alternative problem-solving processes. They are operating components of a more general process, and can serve evolutionary experimentation in two ways.

First, they may be utilized in producing partial or subsystem solutions within the limited domains where they can be sucessfully applied. The fact that these approaches are available to us in restricted spheres where our acquired information is particularly good, allows us to reduce the difficulty of human problem solving by factoring out parts that are susceptible to their more deterministic approach. These partial solutions can then be incorporated into the more judgmental and experimental aspects of social problem solving.

For example, a small enough and simple enough social system like a manufacturing enterprise may, over a limited time span, be able to employ the more deterministic concepts with good effect in enterprise management. It may on occasion experience simple growth where the scale of its operation may increase without employing any modes of behavior novel to the system. Such a change of scale is a kind of natural increase. It can only occur when there is excess capacity in the system and the social requirement it fulfills is either not fully satisfied or is expandable over time. This would imply an opportunity for management to effect changes in the aggregate scale of the system performance. Because of the allometric relationships that link component scales to system scale, it would also imply the necessity for management to alter the mix of its components.

Under limited circumstances, and where effectively centralized control exists, management may also introduce a new technique to the enterprise through a process of programmed learning. This can only be accomplished when individuals and component subsystems identify strongly with the social system goals and subordinate their individual behaviors to them—or, alternatively, when individual and social system goals are fully consistent. If they are willing to submit to passive reprogramming and willing to act as if they were mechanistic components of the larger system, this mode of prediction and planning may be successfully applied. Circumstances do occur where this kind of planning behavior can be applied in relatively simple social systems.

Second, the less general concept models may play a useful role even when, over longer periods of time and for more complex social systems, the pretensions of an approach that sees planning as "engineering the prophecy" must break down. In this role, also, they can sometimes serve as useful instruments in the process of evolutionary experimentation.

The interest in prediction continues. The various techniques of econometric and demographic projection are still employed, but this use becomes transmuted under the influence of the more general paradigm. Such projections are no longer viewed as templates for system engineering, but as instruments in probing realistic options. They are employed to develop some sense of the direction and perhaps the potential rates of change in certain dimensions of the system. They are not taken as an anchor for the future system design but as a rough indicator of emergent problems and as a finger of light pointing out some aspects of the future terrain—but becoming dimmer and less reliable as it approaches the time horizon.[10]

---

[10] Paradoxically, this change in the point of view is encouraging to those who have an interest in forecasting techniques and structural analysis. It reduces the requirements for their relevant pursuit. If one is interested in identifying the components of functional structure and the logic of bundled objectives and transfers in sufficient detail to permit the design of an optimum terminal-state system, one is placed under a discouraging if not impossible burden. No part of that information is

145

In such a role, the more deterministic models can aid evolutionary experimentation in identifying problems and behavioral options. For example, if the fulfillment of the goals of a social system can be viewed as the result of the operation of a simple deterministic process, the deviations from successful goal satisfaction and the possible sources of these deviations may be highlighted when the real world process is compared with the idealized model. It may, thus, play a role in the identification of anomalies leading to a developmental hypothesis.

Furthermore, in the formulation of a developmental hypothesis one may gain some insight into the comparative merit of behavioral options if one utilizes deterministic or game behavioristic models to perform limited pretests of the options. In this fashion one option may be chosen for a more realistic test through social action.

Still further, in carrying out a program of social action that introduces novel behavior, programmed learning may prove an essential and useful tool. An innovation like the computer may require the training of a work force of programmers, and the model and procedure for their training may conform closely to that of the machine system open to fixed-purpose reprogramming.

In short, the less general models of social change may survive in the service of the more general processes of evolutionary experimentation.

THE BIOLOGICAL ANALOGUE REVISITED

We can now perceive that one of the dominant features of the process of biological evolution is manifest in the social process, even though it operates at a different level and in a different way. Recall

---

useful until a sufficiently complete set has been accumulated to allow the design of the optimum total-system solution. Similarly, the exercise of forecasting is no longer required to support a burden of prophecy sufficiently firm to identify the scale of parameters essential to an optimum solution. In the context of evolutionary experimentation these kinds of information commence to support a better understanding of problems and a better formulation of conditional developmental hypotheses in the form of social policy. These applications do not require information to be complete before it becomes useful. Of course, the more this kind of research can inform the learning process the better, but it commences to serve with the beginning of the effort.

the fact that the evolutionary process manifests itself sometimes as adaptive specialization and sometimes as adaptive generalization. Whether the phylogenetic process leads to one or the other depends upon certain aspects of the history of the species and the environmental context in which it is operating.

In the comparison of biological evolution and social evolution made in chapter III, we concluded that adaptive specialization as the behavioral specialization of whole independent populations was less important in social evolution than in biological evolution. Specialization showed up as component specialization of interdependent groups rather than homogeneous specialized behavior of an independent population, as in biological evolution. When the results of social evolution upon social populations were compared in a manner directly analogous to the results of biological evolution upon biological populations, we concluded that adaptive generalization was more common in the social process and adaptive specialization more common in the biological process.

However, when the focus is upon the process itself rather than the effect of the process upon independent populations, we can see that the process of social evolution manifests itself sometimes as adaptive specialization and sometimes as adaptive generalization in a manner similar to phylogenesis. Normal problem solving under the control of an established set of system goals and an established image of the system entity exhibits a *process kinship with adaptive specialization.* Extraordinary problem solving that leads to changes in system entity or goals—a paradigm shift—exhibits a *process kinship with adaptive generalization.* Both are manifestations of social learning as evolutionary experimentation. As in the case of phylogenesis, which form the process takes at any time depends upon certain aspects of the history of the system and the environmental context in which the system is operating. As in the case of phylogenesis, normal problem solving is more common than the more general adaptations leading to paradigm shifts or social reorganization. Although in this and the preceding chapter evolutionary experimentation has been shown to be a process emergent from and fundamentally distinct from phylogenesis, we can see in the interplay of normal problem solving and paradigm shift another instance of process similarity.

147

It may be useful to carry this line of thought one step further, using the generalizations of Simpson (1953) concerning the preconditions for adaptive generalization in phylogenesis. In chapter II (p. 54) these were identified as follows: (1) A new adaptive zone is available, which means that the species must have geographical and ecological (functional) access to it. It is to a degree preadapted, which is another way of saying that the history of the biotype and the history of its environment have created a threshold of opportunity. (2) The species must have a variegated genetic pool not excessively narrowed by biological specialization. (3) The organisms in the species should possess some relatively undifferentiated tissue that can be absorbed in the service of new structures and functions.

To paraphrase these conditions: the greater the adaptability or learning capacity of a species and its organisms, and the more frequent and proximate the threshold of opportunity generated jointly by the history of the species and its environment, the greater is the likelihood that phylogenesis will yield adaptive generalization.

A similar set of statements can be made about the circumstances that favor a paradigm shift or social reorganization as the outcome of evolutionary experimentation. Certainly, the greater the learning capacity of a social system the greater is its potential for social reorganization. Changes in social entity and social goals always require a larger learning effort than functional refinement of the system. As in the case of phylogenesis, this learning capacity is a joint function of the learning capacity of the individual and the learning capacity of the group.

Taking the individual as a focus, there is a parallel in sociogenesis to Simpson's "undifferentiated tissue of the organism": we are well aware that major social changes are often initiated by the young. Indeed, those who deal with the problems of depressed areas are sensitive to the problems of development presented when the age distribution of the population becomes heavily weighted with elderly people. It is the young that possess that degree of "undifferentiated and plastic tissue" that favors social learning. Beyond that, of course, since human behavior is acquired and human organisms can even learn to learn, in sociogenesis this learning capacity is capable of cumulative improvement and even conscious modification.

Similar generalizations can be made when we consider learning capacity at the level of the social system. The more complex a social system and the more variegated its behavior, the more likely it is to generate social reorganization and revised objectives. It has the behavioral variety that facilitates the formulation of new developmental hypotheses and their test through social action. It already has acquired experience in hierarchical organization of systems of control that makes it easier to visualize and implement the development of a novel functional layer. We continuously witness the evidence that supports these propositions. It is well known that a Boston is more resilient than a Pittsburgh in the face of social change: it is clear that the more advanced and complex nations in the Western world experience a more rapid rate of social change and reorganization.

Similarly, the threshold of opportunity presented by the state of the history of the social system and its environment is significant. The problems and opportunities that exist in an exogenous environment become amenable to control through the development of a new layer of control ("internalizing the externalities" in economic terms): the endogenous environment that offers a new opportunity amenable to control through the "spin-off" of a newly defined subsystem. Both present the social system with the chance to gain through reorganization (i.e., through the redefinition of the boundaries that shape its entity). The greater the frequency of these opportunities, the more likely is evolutionary experimentation to generate social reorganization. The proximity or viability of these conceptual and behavioral shifts is, itself, conditioned by the process. As we have seen, persistent failure to overcome the anomalies that prevent the attainment of social goals by means of normal piecemeal problem-solving brings about a recognition of the opportunities presented by social reorganization.

This review underscores an important point. These so-called "preconditions" for adaptive generalization or social reorganization do not form the basis for positive prediction that a social system or biological species will at a specific point undergo major changes in form and function. Precise system states and clearly defined thresholds of a concrete sort are not offered. The generalizations concern the nature of the process. The operation of the process itself periodically generates conditions that favor functional shifts. It is the state of the system with

149

reference to its capability for applying the learning process that is most significant.

This is important to understand because it means that there are no precise, invariant, deterministic functional or organizational preconditions to development such as we have been prone to search for in our fascination with stage theories of development. The principal problems and opportunities with which we are faced in both the developed and underdeveloped worlds do require social reorganization and paradigm shifts to be successfully negotiated. But the foregoing suggests that the social action that will make the greatest contribution to this end will be that directed to improving the capacity for social learning at both the individual and social system levels.

The Orthogenetic Fallacy in Prediction and Planning

It is possible now to draw another rough parallel between social evolution and biological evolution—one that throws into high relief some of the errors man makes in attempting to solve social problems and realize social opportunities.

It seems that we are witnessing a widespread commitment in social prediction and planning to something like the concept of orthogenesis in biological science. The reader will recall that in biology orthogenesis is a doctrine which interprets the major structural changes and the trends established by the large-scale features of the evolutionary process as being caused by an "orthogenetic principle" that operates independently of the process of environmental selection. The principle is nearly always affirmed to have a goal. The major modification in whole biological systems is presumed to be the product of either an "inner urge" (e.g., Bergson's *élan vitale*) or the product of some non-natural external agent. Such a doctrine has been widely discredited in biology because it rests upon metaphysical statements not subject to scientific falsification. The advance of biological science has generated a series of explanations for the phenomena previously thought to be unexplainable save for the untestable orthogenetic hypothesis.

Modern economic and social science literature does not, apparently, contain an explicit doctrine of orthogenesis. It can be claimed, however, that the concepts behind a great deal of social policy formulation are based upon an implicit presupposition of something very much like the orthogenetic principle. When this presupposition is present in extreme form it is readily identified and rejected by social scientists who, however, often reintroduce it in more subtle form.

In its extreme form it has been around for a long time. It says that there is a terminal state of society that is ordained by a metaphysical god or by history. Such an ideal state can be generated *whole* through the intervention of an external agent. At first this concept took the form of the religious millenium. God has a utopian plan for our future that he will establish when the time is right or we have proven ourselves. Since such a state could not be imagined on earth it was projected into the metaphysical realm; later, it became a prospect for the future of society on earth.

The advent of the era of science modified this view. Man became more knowledgeable about physical and historical processes and more skeptical about metaphysical processes. He continued the orthogenetic presupposition, but he removed God and substituted man. It was not a complete substitution, however, because man was more willing to take on a measure of God's power than his responsibility. Those few adventurous souls who produced their own utopias never seemed to wear the God mantle sufficiently well to be convincing. The more acceptable alternative seemed to be: "Let history define the social goal." Both God and man were absolved of responsibility. The terminal state of society is there to be read from social history interpreted as a deterministic process. The determination is "scientific." Social planning is a process of reading the implications of that terminal state to identify the present human action that will short-circuit the laborious processes of history and bring the historical millenium to rapid fruition. If this plan requires tremendous dislocations and human suffering in the process, there need be no nagging moral doubts because man is the agent of the God of history and science and is assured that the normal historical process would extract an even higher toll. The fact that the God of traditional religion is replaced in later form by the God of science does not modify man's agency relationship.

This historically based and supposedly scientifically oriented form

151

of the orthogenetic presupposition has received widespread acceptance in the communist movement. "Scientific materialism" is a major force in the world.

The professional social science community has reacted against this view of the social process and has produced various versions of an "anti-communist manifesto." In the main, however, the criticism has not been directed at the most vulnerable point of the movement—its orthogenetic presupposition. Indeed, the same presupposition has been retained in modified form by segments of the latter-day social science community. (Popper, 1964, is good on this subject.)

The modification is directed more to removing what is thought to be the most objectionable feature of the communist movement—its teleological character (in the sense of hypothesizing a final cause or terminal state). Modern science rejects the notion that the terminal state of any physical or social system can be read from the course of history and would classify all such statements as metaphysical. However, it is sometimes assumed that planning may still take its goal from history if it is seen as an intermediate goal determined by more objective and justifiable scientific methods. This view sees history as an arrow that projects into the future. The limitations on the range of vision prevent perceiving its terminus, but over some reasonable range of time its direction and orientation can be ascertained. It is assumed, therefore, that we can preadapt by designing the social system that some projected intermediate future point will require, and undertaking to assist history in generating the transformation. Through prediction and planning man contributes to the efficiency of the historical process and facilitates its ends.

For all their rejection of a terminal state teleology and the paraphernalia of "scientific" prophecy, such approaches still contain a hidden orthogenetic presupposition *if* prediction and planning are approached with the assumption that (1) the historical process is deterministic and planning is a process of reactive preadaptation; and (2) detailed social engineering can be undertaken for large-scale, complex social systems. The argument of this chapter is that this is far beyond the technological capacity of science. This view of prediction and planning implicit in deterministic growth and development concepts is the disguised, social science equivalent of the doctrine of orthogenesis.

152

When the biologist was confronted with large-scale changes in whole biological systems that he could not explain, he hypothesized an orthogenetic principle or metaphysical agent standing outside the natural process. These major structural changes were later demonstrated to be the effect of the regular stepwise adaptations of the phylogenetic process taking place under special conditions. When man as a social analyst and social activist perceives the desirability of major structural changes in social systems, he is tempted (because of his limited understanding of the social process and a natural impatience to experience the millenium in his lifetime) to assume that the historical process can somehow be short-circuited by man. He assumes that man has the power to stand outside the process as a sort of non-natural agent and direct the system to a predetermined state. He, in fact, applies the orthogenetic fallacy to social prediction and planning.

But man is in no position to stand outside the process of social learning in this sense. The social process is different in several significant respects from phylogenesis, but it does not differ in this respect. It is still an evolutionary, experimental, cumulative process and we are constrained from taking "giant steps" by our intellect, our information, and the continuity requirement.

Under the influence of the orthogenetic fallacy man has often assumed the power of historicist prophecy and the holistic application of social system engineering. In fact the reverse is true.[11] Man is not the midwife of history; he is her consort. History is the record of social learning—the offspring of the ideas man plants in her womb. Just as macromutations in the biological realm are lethal to the offspring, the

---

[11] In this connection the reader is encouraged to read section 21 of Popper (1964). The reader can also find an extensive treatment of the development of historicist points of view in intellectual history and how it has influenced modern social science in Becker (1968). The following statement by Becker (p. 62) is suggestive. "The idea of progress was . . . 'off-centered' in several different ways and in several different countries—in England by Malthus, Spencer, and Darwin, and in Germany by Hegel. In France, Comte . . . contributed to this off-centering by his own reliance on history, by objectifying it in his system. This gave 'scientific historicism' wide currency and the rest of the nineteenth century, as we know, went ahead full steam to try to find 'objective' laws of historical progress. Ethnologists, sociologists, historians—all looked to reconstruct the stages of development of humanity in a single line. This would give them an easy key to the secret of social dynamics."

discontinuous approach to social problem solving, which is implicit in deterministic paradigms, leads to abortive results.

Many may be plagued by the same kind of difficulty that disturbed students of biological evolution for so many years. They might be heard to say, "Evolutionary experimentation may be adequate to bring about normal incremental changes in social behavior, but it cannot account for the occurrence of major social reorganizations. Nor does it constitute a technique for bringing about the major social changes that are needed." In an earlier day the students of biological evolution made a similar claim that phylogenesis could not explain structural and functional novelties or adaptive generalizations.

But the thing we learn from Kuhn (not from Popper) is that the normal practice of evolutionary experimentation is the source from which the major changes in social form and function arise, just as adaptive generalization is another manifestation of the process of biological adaptation. Rather than attempt to bypass the process, we should focus our attention on understanding it better in order to improve its efficiency.

The Efficiency of Evolutionary Experimentation

At this point one might legitimately ask, "Is not evolutionary experimentation a hazardous and haphazard way to invent history?"

It certainly has been a hazardous and haphazard process throughout the entire course of human history. Throughout this time span evolutionary experimentation has been dominated by normal problem solving that was ex post or reactive in character. Man confronts a problem, he hypothesizes a solution, he tests it in action. Through his action he modifies social environments both directly and indirectly, creating new problems while he resolves the old. Both biological and human history attest to the traps that have been generated in this way. Sometimes the new problems that are created through environmental change are less amenable to solution than the old, and lead to extinction or an isolated, stagnant, static state. Man sometimes unwittingly lets the genie out of the bottle.

As we move into the modern era, man's emerging understanding of the process of learning combined with his accumulated knowledge of many natural processes have enabled him to organize some components of the learning process and direct them to the solution of anticipated as well as experienced problems and opportunities. He has also gained a limited power to act upon the environment and to conceptualize and solve problems in the context of larger and more complex systems. This has vastly accelerated the process of social learning and has yielded the solution to countless problems, but the hazardous nature of the process may have been augmented instead of diminished.

There are two interrelated reasons why the hazards may have increased. The first was touched on in the previous section. As man has grown more successful in imposing his will upon the design of deterministic physical processes, he has come increasingly to suppose that he can impose his will upon social processes in the same way. He often develops the pretension that he can engineer social systems just as he engineers physical systems. This supposition has led him to attempt to bypass evolutionary experimentation—to short-circuit the process of social learning. If he experienced an existential condition deemed favorable, he often sought to stop history in its tracks. Conservatives throughout all ages have wanted to limit social learning to the functions of social maintenance. "Stop the world" has been their motto. If man experienced an existential condition deemed unfavorable, he often sought to engineer some "ideal" future state. "Come the revolution" has been the corresponding expression of faith of the radicals throughout all ages.

But the process of social learning will not be stopped. Attempts to stop it or engineer it may sometimes raise hazards to such a peak that the response of the disillusioned becomes "Stop the world; I want to get off." Not recognizing the hypothetical and conditional character of social laws, man has tried to apply spurious social laws to system design in the same manner that he would design physical systems. In the attempt, he violates what may be the only genuine social laws—the laws of process themselves.

The second reason why the process may have become more hazardous was outlined at the end of chapter III. The application of the classical scientific method (and its activist counterpart—systems engineering) has generated the exploding advance of physical system design and physical environmental control. But the effect of this

155

progress has been to require new social forms that support their use. We have been forced into a period of rapid social reorganization, but that reorganization is still following a reactive, ex post, problem-solving pattern that is not adequate to carry out the "forced draft" rate of social learning without severe social dislocations. Social learning is not keeping pace with the rate of social change. Man is not learning effectively from his experience because he is not approaching experience from a learning point of view.

A statement by Nelson et al. (1967, pp. 173–74) illustrates the haphazard and inefficient way that social learning takes place in the formulation and initiation of public policy even in as scientifically and technologically advanced a society as the United States. In referring to one of the range of social problems with which public decision units are currently attempting to cope they state:

Under such circumstances the most fruitful way to proceed is sequentially and experimentally; neither by doing nothing because knowledge is less than perfect nor leaping farther than necessary in a pre-judged direction. Government policy making presently has a tendency to vacillate between these extremes. There is a tendency to delay for a long time the introduction of a new program because of uncertainties and then suddenly to jump in fully with a large commitment to a prescribed program, with no better knowledge base than before, when political pressures for doing something became strong. Once proposed or initiated, the program is then popularized among the public and in the Congress as a sure antidote, rather than as a promising probe of the environment.

This knowledge myth, which forces dedicated public servants to engage in charlatanism, seriously impedes the development of public policy. The channeling of large sums of money into programs predetermined on the basis of sketchy information narrows the range of alternatives that can be tried, and thus reduces the range of policy instruments that have to be tested. Further, it deters useful experimentation, since all programs are action programs. It places a high premium on actions likely to yield simple-minded quantitative indexes of immediate success. . . .

Conditions of great uncertainty call for imaginative and flexible probings, not vacillation between inaction and commitment.

In short, where the learning process is not understood and consciously applied, adaptation to new problem situations that are complex tends to vacillate between the paralysis of inaction and the temptation of solutions visualized, or at least merchandized, as a holistic panacea. We can recognize these oscillating behavioral responses in individuals as well as social systems of all levels. Where policy is formulated and implemented it is rarely consciously and deliberately designed and carried out in ways that will generate the information necessary to support objective testing and facilitate the progressive revision of policy. In the absence of such procedures we often fail to correct mistakes and persist in behavior that aggravates social problems and exacerbates human misery. We have all witnessed the tragedy of individuals and social systems that seem to lack the capacity to learn from their mistakes.

If the process of evolutionary experimentation *is* hazardous and frequently haphazard, as has been stated, is there a better way to invent history? This is the insistent question. Man is posing this question in a rising crescendo as he faces new threats so large that he can scarcely conceive them, issuing from origins so small—the nucleus of the atom, the molecule of DNA—that he can scarcely perceive them.

How, then, can we reduce the hazards of this historical process? The answer seems clear. (1) We must put aside the pretension that the process can be short-circuited. Historicist, social engineering pretensions must be understood as a trap. (2) We must come to understand the process of evolutionary experimentation in the progressive redesign of social systems so that we can make a more conscious and purposive application of it to match our powers of social learning in the domain of physical systems. The hazards of evolutionary experimentation will be diminished if we can make it less haphazard. To attain this result we must find ways to (a) improve the efficiency of evolutionary experimentation and (b) improve its directed character.

First let us deal with point (a). Where man has succeeded in the conscious application of evolutionary experimentation to social problem solving it has been primarily in the area of normal problem solving that could be carried on under the control of established social system entities and goals. But the most critical development problems the world faces today are those that require paradigm shifts or progressive social reorganization. This means that a high priority should be assigned to research that yields greater insight into the character of

157

social learning. We need to begin immediately to apply the emergent insights to promoting those conditions that tend to favor adaptive generalization or social reorganization. Education will need to be addressed more to expanding the learning capacity of individuals and their understanding of the learning process. At the social system level institutional forms will need to be developed that favor social experimentation that works progressively to redefine system boundaries and goals.

As to point (b): throughout social history the problem-solving technique of social learning has been a directed process, the nature of which is capable of improvement. Idea mutation or new behavioral hypotheses are not random as in phylogenesis, but are directed by the generalizing capacity of the human mind to the explicit solution of an explicit problem. The difficulty is that it is common for the direction or purpose or goal to be taken from the problem. The perceived dimension of the problem and its shape, and therefore the nature of the solution experiments, are formed by the biological and acquired social values of the individual or the social subsystem forming the response. But with a fractionated set of social purposes forming a diverse pattern of adaptive responses to environmental problems, responses are often formed that are inconsistent because of differences in social goals.

The means are frequently lacking to formulate agreement on goals at a social system level appropriate to the dimension of the problem and the requirements of its solution. To some extent these inconsistencies tend to resolve through a series of iterative and competitive co-adaptations that are deviation counteracting in nature. Even where this takes place the struggle extracts a toll in human suffering. But in the social realm the aspiration to impose the system's will upon the environment and the limited capacity to do so, frequently lead conflicting systems into an attempt to impose their own goals upon each other. All too frequently these attempts lead to deviation-amplifying adaptations creating explosive situations with magnified human suffering. In the process man has created new problems of frightening dimension for which he has yet found no solution.

Such goal conflicts can only be resolved by goal modification or by subordinating the goals in conflict to a higher order goal that embraces both. Even without conflict, goals may need revision in the light of the results of evolutionary experimentation. But the role of goals and values has received little attention from those who carry out social science primarily under the control of deterministic or stochastic models. When one adopts a social learning metaphor, the significance of these goals and values is inescapable.

V The Role of Social Goals and Values

IN CHAPTER III WE IDENTIFIED those unique elements of the process of social evolution that differentiate it from biological evolution. All of these elements are derived from the uniquely social and conscious cognitive process of the human organism so that conscious behavioral adaptation comes to assume dominance over genetic modification. Among the elements of uniqueness is the prevalence of behavior directed to changing behavior. The last chapter was devoted to expanding upon this characteristic of the social process. It was concluded that this mode of social behavior must take the form of evolutionary experimentation save for those limited instances where social systems approximate homeostatic deterministic systems.

Since the other elements of uniqueness come into play as aspects of evolutionary experimentation, it is not surprising that they have already been encountered in that discussion. For example, the nature of evolutionary experimentation could not be discussed without talking about the goals or objectives of the social system. We learned that the system goals help shape the context within which evolutionary experimentation is carried out and that they form the test of developmental hypotheses.

Of necessity, behavior directed to changing behavior embraces behavior directed to evaluating behavior. Indeed, when we considered how the hazards that have attended social history might be ameliorated, it appeared that one way might involve improving the process of evaluation or the directed nature of evolutionary experimentation. It follows, therefore, that if a social science is to concentrate on evolu-

160

tionary experimentation or social learning and establish principles for its sound practice, it must become involved with the role of goals and values.[1]

The Resistance of the Scientific Community

The scientific community has traditionally resisted the notion that it should deal explicitly with social values and goals as a part of its enterprise. There has been a strong belief that the domain of values and goals is "off-limits" to science. In part this has been due to epistemological presuppositions that relegated values and goals to a subjective realm not accessible to scientific inquiry. The tradition of logical positivism, so strong in science, has held that judgments of value are merely emotional or verbal assertions removed from categories of truth and falsity—a position also characteristic of some Marxist and linguistic philosophers.

Two other factors were probably equally important. First, in that broad domain of physical concepts where a machine system paradigm is operational, science could make great advances without considering social goals and values. Taking these to be exogenously given, it could restrict its attention to operational rules or efficiency criteria. Only these "instrumental" goals or criteria were considered to be objectively derived from tested physical principles. Second, there was a strong desire on the part of the science community to protect its life and growth from persecution by religious and political institutions that considered the realm of value their exclusive domain. It is a hangover from those uneasy early days when science worked out an unwritten truce with the political and religious establishment.

We economists and other social scientists have been inclined to be hypersensitive and defensive about our claim to a position in science. We have, accordingly, slavishly emulated the physical sciences. We have worked hard to apply to social phenomena the machine system

---

[1] The use of the term "value" in this discussion is not restricted to the special meaning it receives in economics, i.e., the marketable price of objects or services. The term is used here to refer to those values that are a reflection of needs, desires or requirements of living systems. Goals are these values made manifest as guides to and motives for action.

161

concepts that have proven so fruitful in application to many physical phenomena. Successful applied social sciences have emerged where certain social phenomena can be reasonably approximated with machine system concepts (especially in economics). Limited success in some areas has encouraged us to continue applying the same concepts persistently and mistakenly to social phenomena that are manifestations of a process of social learning. We have also been conscientious in the extreme in protecting our "scientific purity" by excluding issues of value as matters beyond legitimate concern of the objective social scientist.

This sense of professional insecurity may have reinforced, in turn, our attachment to deterministic models. The classical market equilibrium model of the economist, for example, is essentially the model of a machine system closed to new knowledge. In such a concept the economic system can be seen as occupied with purely instrumental goals (like producing bread or refrigerators) that are indirectly related to more general social goals considered as given. Under the control of such a model the economist can absorb himself with considerations of system efficiency yielding a static state optimum. He can occupy himself with the instrumental goals and criteria that serve to maintain homeostasis or conserve thermodynamic throughput. It is a convenient way of maintaining the appearance of value neutrality because the instrumental goals and values can be reduced to "scientifically established efficiency criteria."

When he moves to a consideration of the phenomena of social development and is confronted with options for changing social behavior, the social scientist has greater difficulty in dodging the normative problems. Nevertheless, an attempt is made to carry the focus on instrumental efficiency over into the developmental realm. This may help explain the great appeal of the deterministic developmental models. Under the control of the concepts, a change in social behavior can be evaluated in terms of whether or not it efficiently contributes to the transformation leading to a predetermined terminal state. Even those who are concerned with stochastic models (e.g., game theory, decision theory, etc.) are primarily occupied with establishing the rule for efficient procedure in advancing or protecting established goals in the context of uncertainty and competition between the decision units. The nature of the goal and its transformation over time is not seen, itself, to be a function of the process.

Under the control of the social learning metaphor the goals must be acknowledged to be a function of the process. In evolutionary experimentation it is the goals that form the test of the developmental hypothesis, and through its practice the goals themselves are brought under periodic review and modification. If social science is to concern itself at all with the study of this process it will be brought to an inescapable confrontation with the realm of values.

The Evolution of Social Values

We can only agree that values are phenomena that lie outside the domain of science if we assume either (1) that all values are derived from a transcendental source by a process of spiritual or transpsychological osmosis, or (2) that all values are the exclusive product of a subjective psychic process presumed to be inaccessible to science. Neither assumption seems acceptable. We shall return shortly to the issue of the scientific study of values, but first we shall consider their natural origin. The thesis presented here is that human values, both individual and social, have emerged as a product of the evolutionary process. Unless we are to deny that the process is a legitimate object of scientific study, we cannot exclude the role of values as a legitimate and necessary concern of social science.

Human behavior is almost always purposive. It is aimed at accomplishing something if it is nothing more than filling the stomach, basking in the sun or acting in some similar way to reduce the tension set up by the physical requirements of the biological organism. In short, behavior is value-fulfilling behavior.

But in this respect human behavior is no different from the behavior of any biological organism. In a very real sense the origin of values can be traced back into the phylogenetic process. Indeed, we can interpret the biological genotype in part as a set of phylogenetically derived values. Each biological organism constituted an experimental system to be tested against the goal of survival. Those that were not "falsified" by the test carried their successful function and structures into the modal genotype of their species. But they carried something

163

else. They carried a genetically determined capacity for physiological tension when the levels of adaptive behavior fell below acceptable thresholds. These are the instinctive drives of the animal organism. The tension of physiological hunger and the drive to reduce the tension through feeding is an example. There appears to be a capacity for affective enjoyment that accompanies tension reduction. Thus, the biological instincts can be interpreted as genetically inscribed values that are the product of the stochastic evolutionary experimentation we know as biological phylogenesis. Because of the differences in the evolutionary history experienced by each specie, each comes to embody a different set of genetically inscribed values. For example, only the chicken and certain related fowl instinctively recognize the shape of the chicken hawk and the value of fleeing from it.

Even at the primitive level of the evolution of values we learn several things of importance. (1) Behavior is not single-valued. There is always a set of instincts such as those manifest in hunger, thirst, the need for oxygen, the sex drive, etc. (2) Multivalued behavior leads to a value hierarchy. Some requirements take precedence over others. For example, the need for oxygen has to be fulfilled before some of the other needs become operative. Individuals derive a schedule of behavior that satisfies these values in a manner consistent with the hierarchical scale. This is characteristic of multivalued behavior at every level.

No attempt is being made here to reduce human values and motivations to biological instincts. The purpose of this discussion is simply to demonstrate that values are a product of the evolutionary process and, as such, their origin can be traced back into the process of phylogenesis itself. At the same time, there is no point in denying that many of the needs and values of the human organism can be traced to its biological heritage. The behaviorists among the psychologists have written extensively about the importance of physiological requirements in human motivation and the role of tension reduction and its associated affective enjoyment in human behavior.

This is not the end of the story for those animal organisms that have developed the capacity for adaptive behavior during the life cycle. For them these genetically inscribed values come to be overlaid by acquired values that are psychologically inscribed. The organism learns that there are several options for reducing any physiological tension such as hunger. Some require less effort to acquire and come to be preferred for that reason. As the organism encounters certain options, the experience tends to beget preference. Tastes emerge. Some ways of satisfying the need come to be preferred above others. Thus, even the basic physiological values are made more multivalued by the emergence of psychological preferences with respect to the mode of satisfaction. These preferences emerge out of the experimental encounter with the life environment. The hierarchy of values becomes more complex. Some of these acquired tastes or values become instrumental values in the service of the prior values represented by the basic physiological requirements.

Nor is this the end of the story for the human species. These values come to be overlaid by another set socially derived.

First, human individuals, like other organisms that possess a degree of behavioral adaptability, develop an overlay of acquired values that are psychologically rather than genetically inscribed. Because of the human's superior cognitive skills these may be less a matter of the habit forming character of successful acts and more the result of deliberate problem solving. The individual is faced with the task of satisfying the physiological requirements in the face of an array of options. There tends to be a more conscious search of the options to identify those acts that are the most efficient value-fulfilling acts. These acts or behavioral modes come to take on instrumental value in serving the target value of the physiological requirements. Because of the ability to communicate acquired knowledge, the human being can acquire pretested behavior and values from others. Some of his values are the product of a kind of programmed learning in which he acquires group values and some a direct result of life experiment. But even the traditional group values are subject to retesting through experience and all of the psychological values are at some stage the product of successful problem solving.

Among the behavioral options available to the human individual are those which amplify his capacity to satisfy his target values through social behavior. He may perceive that one promising potential for fulfilling a need lies beyond his individual power, but may be accessible to several acting in concert. This emerges as a development hy-

165

pothesis: "If I can find several other people with similar needs or target values, I can impute those values to the group and, acting in concert, satisfy those needs more efficiently." The hypothesis is tested in action and confirmed. As the problem-solving skills of the species advance, group action in pursuit of common goals can make an ever larger contribution to the satisfaction of needs through the development of specialization. More complex forms of group behavior emerge.

Thus, social behavior serves to enrich the behavioral options of the individual and amplify the value fulfilling capacity of the individual act. Social behavior comes to have a value attached to it and that value takes its place among the psychologically inscribed values that are instrumental in serving the target values of the individual. Social values emerge.

Something special happens when a social behavioral option is chosen by an individual as a superior means for satisfying an individual target goal. As an individual he may still look upon social behavior as an instrumental means for satisfying the target goal and the social action may be identified with an individual instrumental value. But in order to avail himself of this option, he has to cooperate in the development of and participate in the activities of *a new and higher order behavioral system.* The personal goals that lead him and his associates to cooperative action become the target goals guiding the organization of the social behavioral system. But characteristically there are a number of optional forms of social system organization and social behavior that might serve these goals. Which are realistic options and which are superior options are discovered through a process of evolutionary experimentation through which the social system tests new behavioral modes in relation to the target goals the social system has derived from individual goals.

This is brought about through the practice of evolutionary experimentation at the social system level. Under the control of the image of the group and its joint purpose, normal problem solving leads to the successive refinement of the organization and function of the system. In the process another set of instrumental values emerges in the form of operating criteria and standards that serve the target goals of the group. These become ordered into a value hierarchy in the same way that individual values come to be ordered.

166

As the process of evolutionary experimentation proceeds, the target goals of some social systems come into conflict with the target goals of other social systems so that anomalies develop. Obstacles arise to the realization of the social system target values (and, hence, those of its human clientele) that are not amenable to normal problem solving under the control of the entity and purpose of each system considered separately. The solution to this problem often requires that both social systems cooperate in the development of a higher order social system which now takes on a higher order set of social values and goals. A paradigm shift occurs that redefines social system identity and purpose.

In this fashion, evolutionary experimentation yields an emerging hierarchy of higher order social goals and values and their counterpart behavioral systems.

In short, social values can be understood as a product of the same process of evolutionary experimentation that gives rise to emergent social structures and functions. The greatest proportion of the time the process is engaged in generating new and refined instrumental processes and instrumental values that serve the realization of the target values of the system. At times, however, we arrive at a point where repeated experimentation reveals that further advance can be made only by modifying the goals of the system or incorporating them in a higher order set of goals and a more general system. This makes clear that the development of new social target values is a characteristic of the paradigm shift we encountered earlier. Once a new set of target values and supporting systems is achieved they may be absorbed for some time with the task of evolving a new set of instrumental values through evolutionary experimentation.

As far as it goes, this sketch still leaves unresolved several questions about the evolution and role of values in the social process that will need to be considered shortly, but it is sufficient to indicate that a hierarchy of individual and social values emerges out of the practice of evolutionary experimentation at both the individual and group levels.[2]

[2] In preparing this sketch of the naturalistic evolution of values, the author was assisted by such sources as Waddington (1960), Goodenough (1967), and Burhoe (1967). Through these one can be led to other sources that deal with this topic.

The Scientific Study of Value

We conclude that the reinforcement and reordering of individual human values and the development of social values are a product of the cumulative problem solving experience of mankind. As such it must be a proper and essential subject of scientific study. The emerging social learning paradigm must incorporate an understanding of the role of values and goals and social science must commit itself to their energetic study.

The late Clyde Kluckhohn (1958, pp. 472–73) presented the main points in support of the scientific study of values more than a decade ago. His three points still serve with little modification.

"First, the existence of value judgments does not make behavioral science impossible in principle." He points out that the sociology of knowledge shows how individual and social values influence the selection of facts and the construction of theories, but claims that this does not invalidate the scientific process. He points out that this is "a special case of the proposition that all discourse proceeds from premises and its validity is limited to those premises." The works of Kuhn and the evolutionary epistemologists also make it plain that even the practice of normal science in the most abstract, objective physical domain is not free from scientific relativism and the procedural premises implied by the paradigm consensus of the time.

"Second, values are cultural and psychological facts of a certain type which can be described as objectively as other types of cultural and psychological facts.... The circumstance that at present we meet with difficulties in the description of values is a function of our inexperience."

"Third, it seems abstractly evident that science can say something about instrumental values as appraised in terms of their relative efficiency as a means to designated ends." Indeed, any science that concerns itself at any point with system design (even machine systems), as opposed to pure description, is engaged in the generation of and testing of instrumental values. Furthermore, on the basis of the discussions above we know two things. (1) The target values or goals are an essential element in the development hypotheses by means of which instrumental values are proposed and tested. Since the instrumental values must be tested in terms of their service to the target values, social science must go at least as far as identifying and specifying the target values of social systems. (2) More important, target values of social systems can be reduced to instrumental values and, hence, are equally subject to the same kind of scientific study. We have learned that the target values that govern the experimental development of higher order social systems are instrumental in serving the target values of lower order social systems. The target values and goals of social systems are the product of an instrumental hierarchy that ultimately has its ground in the target values of human individuals.

Kluckhohn adds that, "To speak of 'values' is one way of saying that human behavior is neither random nor solely 'instinctive.'" In short, values are essential to the analysis of a real change process that is not deterministic. They are an essential component of the process of social learning and require scientific study.

The Efficiency of Evolutionary Experimentation

We return now to the issue with which the topic was introduced. At the close of chapter IV we associated social learning with a process of evolutionary experimentation. We observed that, historically, this process has been hazardous and inquired whether our knowledge of the process might suggest some ways in which it could be made less so. In this connection it was pointed out that evolutionary experimentation is a goal-directed process, and a question was posed: "Can the process be made less hazardous by improving the orientation or directed nature of the process?" We must now give serious consideration to that question.

If our sketch of the evolution of social values is an accurate characterization, it suggests that the process of evolutionary experimentation could be made less hazardous if social planning could more adequately take into account the role of values and goals. The process of evolutionary experimentation is something more than a process of testing developmental hypotheses through application. It can also be interpreted as a process of progressively testing the instrumental values implicit in the developmental hypothesis in terms of the target values implicit in the functional entity of the social system. These target

169

values, in turn, constitute a value paradigm that is tested after the manner of all paradigms. Its superiority is tested in competition with competing paradigms (and their social system embodiments) by providing a more successful framework for the testing of instrumental values and by demonstrating a superior line of logical derivation from the accumulated fund of tested experience. (See Kuhn, 1964, for the development of the idea of testing paradigms.)

Historically, the process of evolutionary experimentation has been made more hazardous because of the prevalence of nonfunctional attitudes about the role of values and goals.

As has been discussed, there is the tendency of the science community to deny a role to valuation. Even where science is engaged, as it often is, in the development and testing of instrumental values, there is a preference to speak of "criteria" and "efficiency." While this attitude will allow scientists to engage in a certain amount of so-called "value neutral" testing of instrumental values in system development, it denies them a legitimate role in formulating and testing the paradigm shifts by means of which the target values of the system are enlarged and transformed. The social science community needs to recognize that it is participating in an ongoing process of social system design that involves not only the testing of developmental hypotheses against the standard of established goals, but also the testing of social system goals in the light of their adequacy as planning metaphors.[3]

---

[3] In this connection a statement by Becker (1968, pp. 362–63) is instructive. "In the human sciences the problem of gaining wide loyalty to a paradigm is no different than in any of the other sciences. ... Only, a subtle new factor magnifies the problem immensely, and gives it entirely new proportions: *in the human sciences it is sharpened to an extreme degree, because the agreement cannot be disguised as an objective scientific problem.* That is to say, in the natural and physical sciences, paradigm agreement looks like a disinterested matter of option for an objectively compelling theory. It does not look like an active social and historical (moral) problem. In the human sciences, on the other hand, *the same kind of option for a compelling theory looks unashamedly like a wholly moral option,* because of the frankly moral nature of its subject matter. The physical sciences, when they opt for an attack on reality, also set in motion massive social and institutional changes of a moral nature. Only, there is a difference between this kind of change than that projected by the human sciences: the process of change is indirect; it does not call immediately into play the deep seated reaction to the habitual human world.

170

Also, there is a common tendency to consider the target values of social systems as absolute goals and even moral absolutes. Indeed, we have seen how the emotional attachment to social target values served to intimidate science into restructuring its domain to a context as nearly value-free as possible.

This tendency to convert the target values of social systems into absolutes has several roots. First, the individual who chooses to participate in social behavior as a useful instrument in satisfying individual target values must enter into a social compact with the others joining in the joint behavior. Each has to modify some individual behavioral options in order to satisfy through social action other individual target values. The result is a set of rules that defines the individual rights and obligations with reference to the social system. This creates a special vulnerability. Insofar as another individual in the social complex fails to make the same commitment and abide by the same contract, one loses part or all of what his commitment is supposed to yield. Thus, the social rules and the target values of the system come to be reinforced by strong emotions and, historically, have often been justified by the sanction of authority.

Second, the identification of the individual with the group is often favored by strong psychological motives. Among the powerful psychic attributes of human beings are the need for socialization and the need to attach meaning or purpose to human behavior. Individuals often satisfy these motives by forming a strong emotional attachment to the group. The "in-group"—whether it be state or club—must surely rank high among the "idols" raised by mankind. The less "full-formed" or well integrated the person as a psychic entity, the stronger such attachments are likely to be.

Third, insofar as social organization and social behavior give rise to power élites who can, for a time, rig the social rules and social target values of that subgroup, the subgroup members come to identify the social target values with subgroup values.

Fourth, the orthogenetic fallacy in prediction and planning discussed in the previous chapter also lends itself to this bias. Those who hold to

---

This helps physical science assume the disguise of a spurious kind of detachment since its reality is removed from immediate repercussions on the human realm."

a mechanistic or historicist view of the social process are prone to identify terminal states that serve as absolutes in the service of which all other social values become instrumental.

These different motives often serve to reinforce each other. Under their influence the target goals of the social system often come to be treated as moral absolutes.[4]

When carried to an extreme, this tendency has the effect of temporarily inverting the process of valuation. What in the end are instrumental values, come to be considered target values. What in the end are target values, come to be considered instrumental.

As we have seen, the higher order target values of social systems evolve because they are instrumental in serving the target values of lower order social systems and, ultimately, the individual. Under the influence of the motives discussed above, the value hierarchy of individuals tends to become ordered in such a way as to identify with the target values of the social system. System values become reified. Individuals tend to lose sight of the fact that social values initially emerged as instrumental values in the service of individual target values. Conversely, the social reification leads easily to a view that sees the individual as instrumental in serving the target goals of the social system.

The historical occurrence of this inversion of the process of valuation has been common. However, this phenomenon does not have the power to bring the process of evolutionary experimentation to a permanent halt. It cannot because it assumes that the social system can operate successfully into perpetuity as a closed system serving its own goals. In the face of a changing world environment it cannot. The anomalies that arise sooner or later expose the fallacy of reification. The target goals of social systems lose their inviolate image. Their inadequacy in the service of important human needs is revealed, and the way is opened once again for creative evolutionary experimentation to emerge.

At the same time these tendencies to treat social goals as value absolutes that test the adequacy of all human behavior have radically increased the hazardous nature of evolutionary experimentation and

---

[4] Goodenough (1967) is helpful on some of these points.

increased the human costs associated with it. It leads to a traumatic cycle of social system reification and breakdown. When such motives are prominent it is difficult for evolutionary experimentation to bring about gradual social system change on a part by part basis. Instead, the social system and its value absolutes meet the test of history in toto. The potential for human disruption in such a process is high.

In the last chapter we found that a more widespread understanding of the process of evolutionary experimentation would contribute greatly to increasing the efficiency of its practice or reducing the hazards. It should now be apparent that this must include an expanded awareness of the role of values and goals. The tendency to ignore them or treat them as absolutes serves to sabotage the process itself.

Are There Evolutionary Guides?

If this were the full story of valuation as an aspect of evolutionary experimentation, we would be left vaguely disquieted—as many others have been disturbed by some of the implications of evolutionary theory. Our characterization of the process seems to have destroyed a series of conceptual absolutes. We have been denied the absolute of a "law of evolution" in the sense that it provides us with a terminal state teleology or prophetic trend. We have been denied the absolute of a steady state system that serves perpetual maintenance. We have been denied the reification of social goals or their justification by authority. We seem to be forced as individuals and social systems to face up to the fact that we have to live life as open systems continuously groping and testing. All values and all behavior appear to be relative. We seem to be committed to an unconstrained relativism.

This is discomforting on both intuitive and logical grounds. It seems as though man almost intuitively seeks to ground his behavior upon absolutes. On logical grounds one is led to question the possibility of absolute relativism. This appears to be a semantic contradiction. Indeed, one of the things we learn from a study of the process of evolutionary experimentation is that normal science and normal social behavior are impossible without at least temporarily or conditionally committing oneself to act as if the controlling paradigm were the

173

final authority. Normal science and normal social behavior have to be carried out under the control of a paradigm composed of a conceptual framework in one case and a social system entity and purpose in the other.

In other words, the very practice of evolutionary experimentation by means of which we cope with the problems of life requires that we act *as if* there are absolute system goals and values that form the test. We are forced to admit, of course, that these controlling paradigms do not exist beyond the challenge of social anomaly. This puts us in the position of saying that the logic of the mind and its affective character or capacity for feeling seem to require that we act as if we were committed to absolute goals while the logic of the process of social learning requires that we keep our paradigms or absolute values open to accommodation with a later truth.

This is not the end of the matter. There exists the possibility of a paradigm hierarchy. Kuhn (1964) discusses this. He makes plain that there are subdisciplines in science that operate under the control of their own metaphors within the framework of a still more general paradigm governing the broader discipline. Recognizing this opens the possibility that the periodic alteration of a scientific or social paradigm might be undertaken under the partial control of values provided by a higher order or more general paradigm.

Is it possible that a paradigm shift might be negotiated under the control of some higher order goals and values? Such a possibility would tend to replace the conventional test of competing paradigms set out by Kuhn (i.e., demonstrated superiority in testing instrumental subsystem values and a superior line of logical derivation from the accumulated fund of tested experience). One could then point out that the evolution of novel or higher order social systems (and the consequent paradigm shifts in social goals and values) that results from evolutionary experimentation might possibly be carried on under the control of a still higher order paradigm that supplied the functional absolutes for that purpose.

In social system hierarchies this certainly can and does take place with respect to the tranformation of what might be characterized as subsystems. So long as an important part of the behavior of any system is sublimated to the objectives of a higher order system, its transformation through evolutionary experimentation is subject to the test supplied by the higher order goals of the supersystem. The problem is confronted where no higher order social organization or system control is established. When we are faced, in the process of social problem solving, with the apparent need to extend the social system hierarchy and provide a higher order target goal, in what way are we to judge the options save through the conventional test of the paradigm shift already articulated? Are there any higher order values not yet fully embodied in social forms that might serve as a guide?

In the past man has been tempted to look to the process of evolution itself in pursuit of an answer. We have already examined the effort to read a historicist trend from evolution that could serve as a test or court of final resort in the evaluation of changes in social behavior. We have discussed the fallacy of such a presumption. Even though we rule out any knowledge of historicist trends or terminal states that can serve as a guide to social change, we may still inquire whether there is something inherent in the process of evolutionary experimentation or social learning itself which suggests higher order values that might serve as guides in its conscious practice. Are there some generic characteristics of adaptive generalization or paradigm shifts that can offer a guide?

For example, we know that successful adaptive generalizations always seem to be characterized by increases in the hierarchical complexity of organization and increases in the versatility, adaptability and plasticity of organization behavior. We also know that adaptive generalization appears to take place when certain preconditions are satisfied. It seems, therefore, that these are attributes to be valued and sought through the operation of evolutionary experimentation. Perhaps so. But it should be emphasized that these characteristics are not the target goals of social reorganization. As we have seen, the target goal of a paradigm shift in social system values is provided by the need to eliminate some social problem or anomaly. These characteristics can only be useful as a source of instrumental values that serves the requirements of problem solution. Indeed, the implicit justification of our concern with the study of social learning is primarily the value of the knowledge of that process in supplying instrumental values and procedures for generating and sustaining useful social change. What is useful is defined by the

175

problem. The problem is always in some measure unique to the place, the level of social organization and the history of the social system.

This leaves the problem of relativism only partially resolved. If one is not afraid to move into the realm of philosophical speculation, evolutionary theory may suggest a few more clues about higher order values that might give the process some direction beyond the relativism of immediate problem solving.

The fact that the process of evolutionary experimentation is engaged in problem solving may provide our first clue. The behavior of social systems is progressively modified by testing developmental hypotheses designed to improve the system's performance in relation to its target goals. The social system itself came into being at some point in evolutionary history because its emergent target goals (constituting a paradigm shift) were instrumental in resolving the performance problems of lower order social systems. These social system target goals form an instrumental hierarchy erected upon the base of the target values of the human individual. It must be here that we have to look for our value absolutes or ultimate values if they exist.

The semantic imagery we utilize to describe social behavior may, at times, tend to mislead us. We generally characterize social organization as a hierarchy of social systems erected upon a base of human behavior. We speak of the emergence of "higher order" systems and "higher order" social values. In a sense this is useful imagery because the individual does constitute the social soma and "higher order" systems and values are more general and inclusive. But they are more general or inclusive because they are more successful in encompassing the target values of human constituents. It is easy to lose sight of the fact that "higher orders" remain instrumental. A more accurate basic imagery would see the individual at the apex of the social pyramid, and not at its base. But the pyramid is constructed by an engineering principle peculiar to life. It is built from the top down. Through progressive socialization and problem solving man elevates himself and adds progressively to the satisfaction of his basic goals by evolving a hierarchical substructure of social systems that amplifies his power. Social systems are the substructure and not the superstructure of mankind. Thus, one of the points suggested by the evolutionary paradigm is that we should never lose sight of the fact that social systems are instrumental and that we have to look to the target values of the individual for our basic values.

Some may be inclined to shrink from such a conclusion because it suggests that the ultimate or absolute values of the social process are based upon instinct-like physiological drives that seem to imply a competitive existentialism that reduces the social process to a routine homeostasis based upon the homeostatic drives of the individual. Such an image tends to play down the fact that the satisfaction of these drives is mediated through conscious human action that employs the amplifying power of social action. They are root motives in cooperative social action. One must insist that the biologically determined component of human behavior does, indeed, provide us with some of the target values that should test social behavior at all levels. These basic drives must be accommodated in successful social system design. Social behavior that does not do so proves irrelevant in the end and must be eliminated or modified through the process of evolutionary experimentation. We have not given adequate thought to the implications for social processes of man's rudimentary biological heritage.

However, the idea that human valuation forms the terminal basis for all social valuation is distorted if it is conceived as reducing all valuation to a derivative of rudimentary physiological needs. Human behavior and valuation are not nearly so constrained. The evidence provided by the study of evolution suggests that, through the remarkable process of phylogenesis, man has come to embody certain basic characteristics of the evolutionary process itself in his behavioral motives.

What are the basic characteristics of the evolutionary process and what seems to be the nature of their embodiment in human values and motives? The irreducible absolute value of the evolutionary process is survival. This value has been embedded in the most basic instinctive drives of all animal organisms, including man. The goal of survival stands behind all the other genetically inscribed instincts. They serve survival. An earlier concern with evolutionary theory and its implications for human and social values led to emphasizing the absolute goal of survival. This, in turn, gave rise to what is now known as Social Darwinism—a political philosophy that led to a revulsion because it tended to legitimatize the status quo and the judgment that might makes right. It also led, for a time, to the rejection of evolutionary

177

theory in social science. The interpretation not only tended to violate man's view of his own nature but also failed to account for many facts about social phenomena.

The conclusion is not that survival is not an absolute target value of the life process, but rather that it does not stand alone unmodified, for evolution has taught us something else. Activity directed to mere existence or self-maintenance, while necessary, is not sufficient to assure survival. Phylogenetic specialization has demonstrated this down through the millennia. Adaptive specializations have been directed to improving the efficiency of self-maintenance. It serves the survival goal admirably in a stable environment. But environments are never indefinitely stable. Terminating genetic specializations and routine biological self-maintenance will serve the goal of survival for only a limited time. Over the long run survival is better served by adaptability than by adaptation. This is the reason why the emergent lineages with the longest survival records have involved those species in which the phylogenetic process has inscribed the power of behavioral adaptability. Over the long run, survival is served best by a highly developed learning capacity. At the human level social learning is an essential component of the absolute goal of life.

It would be strange if this biologically acquired faculty of social learning were not, in itself, an important root factor in the values and goals of individuals along with the physiological drives of the animal organism. Some of these, as we saw earlier, take the form of tastes acquired during the life cycle that represent option preferences between the modes of satisfying the biological requirements of the organism. But the psychological motives appear to run deeper than this and seem to be grounded in the human capacity for a socially supported cognitive process.

This conclusion appears to be supported by recent developments in psychological theory. Until quite recently the two most widely held theories of personality development were those in the behaviorist tradition (stimulus-response psychology, etc.) and the psychoanalytic movement. As different as these theories are, they have in common an emphasis upon tension reduction and equilibrium. It is only quite recently in the work of such scholars as Allport (1955), Buhler (1959), and others that a more open psychological theory is beginning to emerge. Buhler underscores the change in the following way (pp. 564–68): "In the fundamental psychoanalytic model, there is only one basic tendency, that is toward *need gratification* or tension reduction. . . . Present day biological theories emphasize the 'spontaneity' of the organism's activity which is due to its built-in energy."

She goes on to say, "These concepts represent a complete revision of the original homeostasis principle which emphasized exclusively the tendency toward equilibrium. It is the original homeostasis principle with which psychoanalysis identified its theory of discharge of tension as the only primary tendency. . . . [We come] to the conclusion that the classical concept of homeostasis has to be replaced by one wider and more flexible than the original concept. . . . homeostasis . . . needs to be refined to cover the tendency to *change* besides the tendency to *maintenance*. . . . there is no room within this system for tension-upholding or even tension increasing processes as primary tendencies. Play and creative work would, according to this system, not yield any pleasure to the individual while they are going on, that is while the activities put the individual under strain. The pleasure would come only in the relaxing phases. . . . Besides the energy needed to maintain life in equilibrium, there is surplus energy. . . . This surplus eneregy . . . disturbs the homeostatic equilibrium as much as deficit does and therefore must be discharged. . . . *not just discharged,* but transformed into a product that, as an outgrowth of the living being's activities, *amplifies* the individual himself."

Allport (1955, pp. 12, 40) makes the same point in a different way:

... the person is not a collection of acts, not simply the locus of acts; the person is the *source* of acts. And activity is not conceived as agitation resulting from pushes by internal or external stimulation. It is purposive. ... every state of the person is pointed in the direction of future possibilities. [There are a number of] aspects of personality that make for inward unity ... which ... we call the proprium.

He goes on to acknowledge that the operational or positivist models of behavior are certainly valid for many human drives—particularly the biological drives for oxygen, food, sex, etc., but adds (pp. 66–68):

... we are dealing here with only half of the problem. While we certainly learn habitual modes of reducing tension ... we also abandon old habits and take risks in searching out new sources of conduct.... Though we want stability we also want variety.... But risk taking and variation are fraught with new and often unavoidable tensions which, however, we scorn to avoid.

... the formula ... for drive reduction seems to break down when motivation is no longer a matter of segmental drives or opportunistic adjustment, but rather partakes of propriate striving.... Propriate striving confers unity upon personality, but it is never the unity of fulfillment, or repose, or of relieved tension.... its goals are, strictly speaking, unattainable.

We are thus driven to the conclusion that motives are of two orders, though in a given instance the orders may fuse. ... there are deficit and growth motives. Deficit motives do, in fact, call for the reduction of tension and the restoration of equilibrium. Growth motives, on the other hand, maintain tension in the interest of distant and often unattainable goals.

The effect of these aspects of human nature is that man strives relentlessly for conviction or meaning. Becker (1968, pp. 171–72) expresses this well:

Man is the only animal who is not "built into" his world instinctually. An animal with an instinctive set of responses suffers limitations because its world is already "ready-made" for it. Evolution has ... sealed the animal firmly into its adaptational mold. Man alone among the animals gradually develops his own perceptual response world by means of imaginative guiding concepts. He is actually, in this way, continually creating his own reality. The advantages and disadvantages here are obvious. The potential world at man's disposal is enormous; other animals seem to be condemned to experience the same world for all time. In the present discussion we will not be concerned with the *advantages* of this potential openness to new perceptions and experience, but rather with its marked *disadvantage*. It is this disadvantage that gives us a major clue to the nature of human striving in general and to the problem of esthetics in particular.

The fact is that an animal who is already pre-equipped with instinctual response patterns is served up, so to speak, with a world that is rich with intrinsic meaning. A portion of nature is eternally alive for a bird and a cat, calling on their instinctive capacities, challenging their energies. The world that has been already prepared for instinctual response is, in other words, a world that carries its own inherent meaningfulness and conviction. But what does an animal do who comes into the world instinctually almost nude? In what particular ways is he to direct his own energies, how is he to seal himself into the life process around him? Furthermore, the matter is complicated if

180

he is equipped with a large, highly convoluted brain and restless, emotional primate energy. The everyday food quest alone cannot answer to his restlessness; the cycle of eat, fight, procreate, and sleep—that absorbs the adult members of other species—has only the barest meaning for man. He has learned to use language and dwells in dreams and concepts, in a past-present-future, a space-time largely of his own creation. The result is that he *brings more* to the world than it is prepared to give to an animal equipped with instincts alone.

And here is where the problem seems to lie. ... the everyday reality that galvanizes all other animals tends to lack conviction for him. In order to render reality meaningful, in order to stimulate his own productive energies, man must bring his own meanings to the world, impart his own sense of conviction. This is the tragic burden as well as the unrivaled creative opportunity, of *Homo poeta*. Man creates his own meaning, and the penalty for failure is what we would expect; if man creates his own life process, his own reactive world, then when he fails to do this sufficiently or well, he edges back from life; this is what we see in the deculturation of certain tribes and peoples and in the psychiatric syndromes known as schizophrenia and depression. Evolution has thus left man with the greatest burden and challenges; he is born, not into a world, but into a "backdrop," that contains the raw materials for his manipulation and for the creation of his own world.

Thus, of major importance among the values of human individuals is the value that attends the human experience of the creative learning process itself—including the creation of meaning.

It follows that what Allport calls the "growth motives"[5] must be among the target values of the individual that social systems ought to be instrumental in serving. It must also follow that social systems that are instrumental in serving these "growth motives" as well as the "deficit motives" are also serving the most fundamental and absolute requirements of the evolutionary process. One might go further and hypothesize that social systems that support the learning or developmental motives of their human constituents may, in the long run, be the only ones that will support social development consistent with social survival and consistent with the development and exercise of the psychic potential of the mass of human individuals.

[5] This is unfortunate terminology in view of a distinction made earlier here between growth and development. In the context of the present book they might more appropriately be characterized as development motives.

It seems safe to say that, to date, historically evolved sets of social systems have not adequately met the test formed by these values except in special and restricted circumstances. There seem to be two principal reasons for this.

The first has to do with the temporal and functional priority of the deficit motives. Self-reproduction and self-maintenance are the irreducible initial conditions of life for biological organisms. It is with the fulfillment of these requirements that the rudimentary biological instincts or deficit motives are associated. Above all else, every biological organism is concerned with the maintenance of thermodynamic throughput and self-replication. It follows that, during the earlier phases of social evolution, the problems to which a largely unconscious evolutionary experimentation was addressed were associated with the satisfaction of the deficit motives. The social systems that emerged were dominated by target goals designedly instrumental in serving the material needs of human populations. Evolutionary experimentation was addressed to sustaining and advancing the achieved levels of material satisfaction.

As a result, the initial social systems were static state in character— as were the predominant human motivations. Individual growth motives were slow in developing for two reasons: (1) For a major part of the population and for a major part of the span of social evolution material output per person was insufficient to allow most individuals to rise above an overwhelming concern with self-maintenance. (2) The social organizations that have been instrumental in serving these deficit motives have tended over long periods of history to be static state systems primarily designed to maintain the homeostasis of a social thermodynamic throughput. Such organizations characteristically provide human constituents with only a limited opportunity to satisfy and to reinforce individual growth motives. Indeed, the effect of such a social organization is to reduce individuals to cultural artifacts serving as social system components in return for the requisites of physical maintenance.

We have observed that, in the recent historical period, the emergence of consciously directed social learning in the domain of science has brought radical and accelerating social changes. Even here, however, the predominant concern has been with displacing thermodynamic constraints that limit the levels of population and material welfare. At the same time, the effect of this more dynamic and creative problem solving has been to give dramatic new opportunities for individual development and a reinforcement of the growth motives of large numbers of individuals. Furthermore, the levels of material welfare in advanced technological societies are rising to the point where the great mass of human constituents needs to direct only a fraction of its psychic and physical energy to the satisfaction of the "deficit motives." For the first time in human history one can perceive the possibility of developing social systems that are more adequately designed for the satisfaction of human growth motives. With this, there is an emergent political demand that the individual's growth values be consciously accommodated by social organization—that evolutionary experimentation be guided by these target goals as well. This phase of social evolution has not yet developed to the point where social-organization or goal-directed problem solving reflect these motives in any widespread way.

This brings us to the second reason why the historically evolved sets of social systems have not adequately met the test formed by these superordinate values. We do not yet adequately understand how to design social systems that satisfy these target values. Virtually all of our historical experience has been with the design of social systems that have as target goals the satisfaction of the deficit motives. To develop social systems that are adequate in the service of the growth motives requires an understanding of the way that human participation in social action stimulates, directs, and supports the development of the creative capacity of individuals. Conversely, we need to understand how the exercise of individual developmental motives serves the process of problem solving at the social system level. In short, we become concerned with the relationship between individual and social system learning.

Individual versus Social System Learning

Enough has been said here to establish the fact that there is a natural isomorphism between individual human development and social system development. In the long run, individual learning and social system learning tend to be mutually supporting or mutually frustrating. Unfortunately, as Seeley (1967, pp. 50–51) points out, "We are at the unhappy stage in the development of our sciences ... when attempts to rationalize social life founder upon the intractabilities of personality organization and when attempts to improve the quality of personality organization come to grief on the brute rock of social organization. It would be going ... too far to say that we know very much about *the relation of social organization to personality organization* and disorganization." (Italics mine.)

Indeed, the social and behavioral sciences have virtually ignored the critical interface between these two levels of human behavior. Swanson (1968), writing in *Science*, emphasizes this neglect. "There are almost as many works on the development of personality as anyone would have time to read. ... Nothing like this amount of attention has been given to socialization. Development processes are believed to be initiated, and primarily to be controlled, by intrapsychic mechanisms. In socialization, the crucial occasions for a change in personality and the forces that primarily guide that change are provided by agents of the social order. Facts like agency and social order are not treated in psychological theories. Rather they are taken to be adventitious to psychic processes and have no theoretical meaning. It might be expected that this area of study would prove of greater interest to sociologists. That has not generally been so. Sociologists tend to be interested in the structural peculiarities of organization ... , but they have little concern for the processes by which these organizations actually shape the personalities of individuals."

As will be indicated in chapter VIII, there has been some movement in the social sciences toward filling this gap. However, in the main we are faced with the fact that the relationship between individual learning and social system learning falls into a crack between established social and behavioral science disciplines and is not being given adequate attention. Adequate attention in this case may very well mean an all-out social science attack on this problem, for there appears to be no area of human understanding more critical to the viability of the development process.[6]

[6] Klausner (1967) has given explicit recognition to this problem of the interface between human development and social development in an article entitled "Links

This review of the role of values in evolutionary experimentation has underscored a point made in chapter III. Social learning as we understand it must embrace both individual learning and the learning that attends the transformation of social systems. We have seen that it is out of the capacity for individual learning that social systems evolved; it is through the instrument of social organization that human behavior is amplified.

But, while the individual learning capacity and individual motives form the origin of social systems, the development of social systems has transformed the environment within which human motives are formed and individual learning takes place. Social history has demonstrated that we can develop social systems that radically amplify the capacity of man to satisfy his deficit motives. In the process we have created a social environment that is often inimicable to the satisfaction of man's growth motives. Man's relentless striving for conviction or meaning is frustrated. He is not adequately supported by his environment in the exercise of creative learning at the cognitive level or creative participation at the affective level. It can be hypothesized that in the long run it is not enough for social systems to amplify man's behavioral capacity. They must take as their proper role the amplification of the individual himself. The higher order goal that is consistent with, and fundamental to, the evolutionary process is the creation of meaning—providing the framework for the development of human potential. This may be the absolute that serves as the ultimate test of evolutionary experimentation—the absolute that is formed by the process.

and Missing Links Between the Sciences of Man" in which he discusses the problems of identifying transformation concepts which link concepts on two or more theoretical levels and can serve as a mechanism to transform an event in one system into an event in another system.

VI   Social Organization and Social Learning

SOCIAL ORGANIZATION and social system behavior are related to social learning in two ways appropriate to our theme. First, we are interested in the effect of social learning on social organization. How does the structure of social systems change as new knowledge emerges that changes the options of social behavior? How has social organization evolved? Second, we are interested in the effect of social organization upon social learning. How does one organize and manage those aspects of behavior directed to changing behavior? In what way are such structures and practices different from those directed to social maintenance? These are the kinds of issues to which we shall turn our attention here.

But first some fundamental concepts should be clarified. As the theme of this book has developed there have been frequent occasions when system concepts are utilized and social systems discussed. We now have reached a point where it is necessary to have a clearer understanding of what a social system is. We need to inquire how it is that an abstract entity like a social system can be perceived and taken as a legitimate unit of study. It is not a subject that has been treated adequately in the literature. In what sense are social systems similar to or different from life systems or machine systems as entity concepts? By describing some of the general characteristics of social system organization, perhaps we may better grasp the relationship between social organization and social learning and, in the process, gain further understanding of some of the special problems that confront social sciences.

186

The Social System as an Entity or Control System

Some people have the intuitive feeling that when one deals with a social system as a unit of observation, one is in some sense dealing with an entity that is less real than those studied in the physical and life sciences. On this topic Campbell (1958, p. 14) observes: "Social groups as entities do not have an epistemological status different from such middle-sized entities as stones and rats, but are apt to be fuzzier, less discrete, less multiply confirmed, *less precisely bounded,* and *only* in this sense less real. The degree of entitivity . . . is a matter for empirical determination rather than a priori decision." (Italics mine.) Campbell leaves no doubt that social systems are a legitimate unit of study—albeit more difficult to measure and observe.[1] Bolstered by this justification, we turn our attention to an elaboration of the abstract idea of the functional social system.

SYSTEM BOUNDARIES AND SYSTEM HIERARCHIES: POLANYI

We can make a good beginning by referring to an excellent article by Polanyi (1968), appearing in *Science.* He makes the argument that life's irreducible structures are the boundary conditions or the system designs that control or harness the laws of inanimate nature, but which are themselves irreducible to those laws. The flavor and substance of his argument can best be given by means of a series of quotations:

If all men were exterminated, this would not affect the laws of inanimate nature. But the production of machines would stop, and not until men arose

[1] In this connection he quotes the following passage from Stuart Rice's *Quantitative Methods in Politics*: "It is as valid for the scientist to speak of the social group as to speak of an ounce of ether, provided he can do something further with the idea. All he knows of either is what he infers through the mediumship of his own or . . . another person's sense impression. . . . When all possible corroboration is secured, belief in the existence of any material thing remains an inference. Belief in the existence of a group relationship is of the same character. It is an inference . . . ."

The reader will note that Campbell has coined a word—entitivity. The word expresses very well the notion that, because they are not precisely bounded, social systems may vary in the degree to which they can be characterized as entities. It is in this context that the word will be used throughout this chapter.

again could machines be formed once more. Some animals can produce tools, but only men can construct machines; machines are human artifacts, made of inanimate material. . . .

The structure of machines and the working of their structure are thus shaped by man, even while their material and the forces that operate them obey the laws of inanimate nature. . . . we harness the laws of nature at work in its material and in its driving force to make them serve our purpose.

This harness is not unbreakable; the structure of the machine, and thus its working can break down. But this will not affect the forces of inanimate nature on which the operation of the machine relied; it merely releases them from the restriction the machine imposed on them. . . .

So the machine as a whole works under the control of two distinct principles. The higher one is the principle of the machine's design, and this harnesses the lower one, which consists in the physical-chemical processes on which the machine relies.

. . . we may borrow a term from physics and describe these . . . restrictions of nature as the imposing of *boundary conditions* on the laws of physics and chemistry. . . .

From machines we pass to living beings . . . . In this light the organism is shown to be, like a machine, a system which works according to two different principles: its structure serves as a boundary condition harnessing the physical-chemical processes by which its organs perform their functions. Thus, this system may be called a system under dual control. . . . A boundary condition is always extraneous to the process which it delimits. . . . their structure cannot be defined in terms of the laws which they harness. . . . Thus the morphology of living things transcends the laws of physics and chemistry. . . .

Hence the existence of dual control in machines and living mechanisms represents a discontinuity between machines and living things on the one hand and inanimate nature on the other hand, so that both machines and living mechanisms are irreducible to the laws of physics and chemistry. . . .

The theory of boundary conditions recognizes the higher levels of life as forming a hierarchy, each level of which relies for its workings on the principles of the levels below it. . . . Each reduces the scope of the one immediately below it by imposing on it a boundary that harnesses it to the service of the next higher level, and this control is transmitted stage by stage, down to the basic inanimate level.

The principles additional to the domain of inanimate nature are the product of an evolution. . . .

. . . the transcendence of atomism by mechanism is reflected in the fact that the presence of a mechanism is not revealed by its physical-chemical topography. We can say the same thing of all higher levels: their description in terms of any lower level does not tell us of their presence. . . .

Thus a boundary condition which harnesses the principles of a lower level in the service of a new, higher level establishes a semantic relation between the two levels. The higher comprehends the workings of the lower and thus forms the meaning of the lower. And as we ascend a hierarchy of boundaries, we reach to ever higher levels of meaning. Our understanding of the whole hierarchic edifice keeps deepening as we move upward from stage to stage.

While Polanyi is writing about the structure and organization of systems, his terminology is not unlike that used in Part III when we were considering *social behavior* directed to evaluating and changing behavior.

This similarity calls our attention to the fact that behavior and structure, the act and its framework, are both system attributes. They are two sides of the social system "coin." In discussing behavior and in discussing structure we are bound to engage in parallel discussions that employ similar concepts and terminology. That this is natural and necessary has been emphasized by Simon (1962, p. 74) when he says "problem solving requires continual translation between state and process descriptions of the same complex reality."

Polanyi says that a system is defined by its boundary. It is recognized by the shape given to its behavior by its boundary. But this boundary is not conceived as an enveloping membrane that defines some sort of physical limit to the systems. Instead, the boundary is conceived as a behavioral design. It is a pattern that constrains the behavior of subsystem components to action modes consistent with the design. In effect, the system boundary becomes identified with the concept of the controlling paradigm or metaphor discussed earlier.

It is a design for behavior that is not reducible to the behavior of its components or subsystems. Nor can it be inferred from knowledge of subsystem structure and behavior. It is transcendent. It organizes component behavior into a composite behavior unique to the boundary condition. Since the same components could be organized into a different composite or different behavioral set by a different set of boundary conditions, no single boundary can be inferred from them.

The boundaries defining systems are hierarchical and emergent. Subsystems come under the control of supersystems that impart to the action of the subsystem a more general meaning or purpose. These more general boundaries emerge through a process of evolution because they yield solutions to life problems. Emergent boundaries are equivalent to system reorganizations, which we have already identified in chapter IV as equivalent to the notion of the paradigm shift.

The last statement also makes clear that in discussing organization we are no more free of the necessity to cope with goals and values than when discussing the other attributes of evolutionary experimentation. It is the boundary or organization design that embodies the meaning or purpose of the system and organizes lower order meanings or purposes in a manner consistent with its control.

We have said that the system boundary is not identified with an enveloping membrane. This does not imply that such a physical boundary may not exist or be evident to the senses. A system boundary organizes a set of system components and physico-chemical attributes of nature into an articulated system. The effect may be to generate a system with a physical configuration recognizable to the visual and tactile senses of the human organism. This is certainly true of physical systems like machine systems and life organisms. Even here, however, the enveloping membrane that constitutes the physical limits of the system does not necessarily define the functional limits of the system— nor can the topography of the physical limits be used as a surrogate for understanding the boundaries of the system (in the sense of the system's functional control and behavioral scope).

When it comes to social systems and to ecosystems in the life sciences, the physical limits may not even be accessible to visual and tactile senses. A social system boundary arranges a set of human components, human artifacts (including machines), and physico-chemical attributes into an articulated behavioral system. This bundling together of subsystem activities in a joint endeavor may have the consequence of bringing into spatial juxtaposition the physical agents that generate these activities so that they reveal a kind of aggregate mass or topography available to the senses. Thus we can readily identify the presence and location of a city or urban system. But its physical limits or membrane are never completely unambiguous or clearly available to the senses. We cannot usefully speak of the boundaries of a social system in this sense. This is partly because the physical entity in this sense is fuzzier and less discrete than in machine systems and life organisms. But more important, like all behavioral systems, the true boundary or definition of the system is linked to its purpose and design. This concept of the boundary of the system should not seem strange to economists because they are used to speaking of an economic system as being defined by its "production function"—i.e., its behavioral design.

NESTED HIERARCHIES VERSUS OVERLAPPING BOUNDARIES

There is another reason why it is not useful to identify social systems in terms of physical limits or boundaries: social systems do not always form a simple formal hierarchy, but commonly overlap in many ways to form exceedingly complex functional patterns. This functional overlapping is not altogether absent from machine systems and life systems, but it is not nearly so dominant a feature and is seldom emphasized in the descriptions of such systems. Such a statement deserves elaboration, but first we should undertake a more careful examination of the structural characteristics of complex social systems.

There are as many different activities subject to combination into behavioral systems as there are behavioral specializations. In an advanced society these are very numerous. If every activity could develop a meaningful, direct functional linkage with every other activity the resulting complexity would be too great for the human faculty to comprehend. There would be no prospect for an analytical social science or meaningful social planning.

But an activity network does not constitute a uniform fiber. No activity can develop direct and identical linkages with every other activity. No activity acts directly upon, or is directly influenced by, all aspects of its human and natural environment simultaneously. No component of culture has the capacity to be involved in the full range of experience accumulated by society. It follows that activities must be

191

arranged in functionally related clusters. These activity clusters are not simple aggregations. They are organized by system boundaries that impose controls upon component behaviors.

As has been emphasized by both Polanyi (1968) and Simon (1962), the arrangement of these activity clusters is hierarchical in character. According to Simon, this ordering or clustering of activities into hierarchical sets yields the possibility that the whole complex system can be decomposed into system components. He states (p. 73) that "most things are only weakly connected with most other things; for a tolerable description of reality only a tiny fraction of all possible interactions needs to be taken into account." Furthermore, ". . . hierarchic systems are usually composed of only a few different kinds of subsystems, in various combinations and arrangements."

This functional decomposition would be simple to achieve if these functional ties always formed simple or formal hierarchies. However, in social systems this is rarely the case.

To illustrate, it is useful to recognize that all specialized activities tend to fall into two generic behavioral types. There are transformation functions and transfer functions. Transformation functions perform all of those activities that cut, shape, anneal, coat, fasten, separate, heat, cool, change molecular structure, alter biological organisms, modify ideas and signals, etc. They change the form and function of some set of physical, biological, or symbolic entities. Transfer functions perform those transfers of material, biological, and informational entities within a space-time grid that are essential to the exchange implied in joint or linked behavior. The transfer function is conceived to include not only the provision and operation of appropriate channels of transfer but also the operation of accumulators or storage buffers (inventories) necessary to the generation of matched and controlled flows over both space and time.[2]

---

[2] This distinction is closely similar to the conventional distinction in economics between production and marketing functions but is a more elemental and precise concept. The production function concept is commonly used as a generalized concept applying to a production system (e.g., the production of ball bearings or bread) that already constitutes a heterogeneous behavioral aggregate. As conventionally conceived, such a production function would apply to a series of activity interrelationships that contained both transfer and transformation processes. The elemental

Every behavioral system, including social systems, is made up of these basic components of effective behavior. Behavioral systems are formed by functionally linking two or more transformation processes by the intermediary of one or more transfer processes.

When these transformations and transfer processes are single purpose in character the result is a simple or formal hierarchy. This can be illustrated by reference to figure 1. To give the discussion some concreteness, figure 1 can be taken as representative of the organization of the steel industry, assuming that all of the transfers utilized single-purpose channels and all transformation processes produced a single output designed to serve a single higher order purpose.

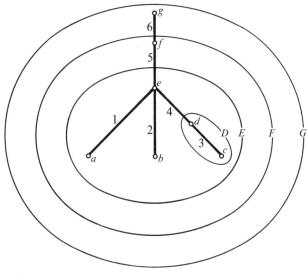

$a \cdots f$ = Transformation processes
$1 \cdots 5$ = Transfer processes
$D \cdots F$ = Nested systems

FIGURE 1. *Illustration of a formal functional hierarchy.*

---

distinction between transformations and transfers is sufficiently basic to apply to biological as well as social processes. Indeed, the performance of biological effectors can be subdivided into the same behavioral elements.

193

Thus, transformation processes *a*, *b*, and *c* could represent the production of iron ore, limestone, and coal, and process *d* represent the production of coke. Transformation processes 1–6 represent single-purpose channels that serve that transfer function alone. They may take the form of conveyor belts, private railways or roadways, pipelines, etc. These transformations and transfers can be seen in various combinations as a nested hierarchy of systems. Thus, system *D* (the production of coke) is composed of subprocesses *c*, *d*, and *3*. System *E* (the production of pig iron) is composed of subsystem *D* in combination with subprocesses *a*, *b*, *1*, and *2*. System *F* (the production of steel) is composed of subsystem *E* plus subprocesses *5* and *f*. System *G* (the production of semifinished steel—i.e., blooms and billets) is composed of subsystem *F* plus subprocesses *6* and *g*.

However, social systems rarely take on this simple, hierarchical nested form because these transformation and transfer processes are often multiple purpose in character. When this is the case, the functional sets that make up social systems do not nest unambiguously. They overlap in an extraordinary variety of ways. This can be given elementary representation by reference to figure 2. The semi-finishing mill, *g*, can be seen sharing its output with a pipe mill, *h*, a rod and wire mill, *i*, and a sheet mill, *j*. Thus, *h*, *i*, and *j* share jointly in the output of *g*, and systems *H*, *I*, and *J* all overlap with each other and with system *G* at terminus *g*. Furthermore, these can all be viewed as part of a larger system *K* where the components systems *H*, *I*, and *J* are joined in a common venture to gain the economies of scale offered by *g* in preference to operating independent and parallel systems.

This is only the beginning. The problem becomes too complex to extend the diagrammatics, but these lines of connection can form complex feedback loops that vastly complicate the system hierarchy. Part of the output of the sheet mill, *j*, may go to an auto manufacturer and a part of the output of the auto manufacturer may feed back as input into the coal or limestone processes—*b* or *c*. Part of the output of the coal process, *c*, might bypass process *e* and go directly as an input into *f*.

Some systems are not only multiple purpose like process *g* (semifinished steel) but some are multiple output in nature like petroleum refining. For both of these reasons a process or subsystem may act as

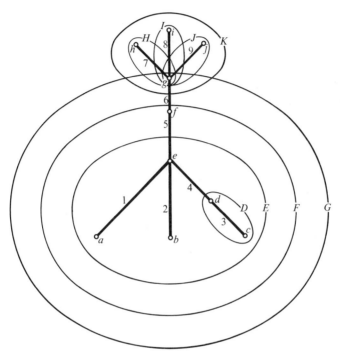

FIGURE 2. *Illustration of functional overlapping resulting from multipurpose activities.*

a component or subsystem on one utilization with respect to only one aspect or a fraction of its behavior. It may simultaneously be a subsystem component of more than one system, and it might be a first-order subsystem with respect to one and a fourth-order subsystem with respect to another. Furthermore, the same activity may constitute, at the same time, both a first- and second-order subsystem relationship to a higher-order system in the same hierarchy. But perhaps the most complicating fact of all is the emergence of general-purpose transfer systems where the widest variety of activity systems come to overlap through the use of a common channel of transfer.

In short, the multiple-purpose nature of many transformation and transfer processes results in a state where the activities performing these functions may *simultaneously* perform as components of two or more higher-order control systems. This leads to overlapping boundaries and reduces the adequacy of the familiar concept of the formal nested control hierarchy.

OVERLAPPING BOUNDARIES AND THE DEGREE OF ENTITIVITY

Several conclusions can be drawn from this representation of the convoluted fabric of a complex social system.

First, it is obvious that complex social systems cannot be identified or defined in terms of observable physical boundaries.

Second, they can be subdivided into systems in many varying ways. Because of the presence of multiple-purpose activities leading to functional overlapping, the widest variety of system boundaries (in the sense of an organizing principle) can be identified in terms of the number and variety of activity components and levels of system hierarchy.

Third, some of these systems will be more clearly defined than others. In short, some exhibit a higher degree of entitivity than others. In overlapping systems the boundary conditions that define a system are rarely the sole operating organizing principle affecting the behavior of component systems. Therefore, the only system that is unambiguously defined is a completely independent self-sufficient system—one where all of the functional linkages are endogenous and under the complete control of the organizing principle or boundary condition. The degree of entitivity is a function of the degree to which the boundary condition defining a system must share the control over its subsystem components with the boundary conditions defining other overlapping parallel or higher-order systems.

This suggests a proposition. The degree of subsystem control exercised by a boundary condition and, therefore, the degree of entitivity of the system, is related to the degree to which the system components retain direct functional linkages to exogenous systems. If the subsystems under the control of a higher order boundary condition retain no exogenous linkages and form a formal hierarchy, the degree of entitivity is very high. The effect of any exogenous links upon subsystem behavior would be an indirect one mediated through the direct control of the boundary condition of the system. Thus, the formal hierarchy depicted in figure 1 is a system with a high degree of entitivity. Such a system is only one step removed from a completely independent, closed system (without exogenous links) that represents the conceptual limiting case.

In those cases where system overlapping is present the degree of entitivity is necessarily lower because a system must share the control over subsystem behavior with at least one other exogenous system. A subsystem is subject to the control of at least two boundary conditions each operating with respect to a portion of its behavior. In these cases the degree of entitivity may be said to be proportional to the richness of connection the subsystems retain that are under the direct control of exogenous systems. Expressed another way, it is the ratio of the endogenous linkages mediated directly by the boundary conditions of the system and the exogenous linkages that link subsystems directly with external systems not subject to the direct control of the given system boundary.[3]

[3] This definition of the entity of a system and this measure of the degree of entitivity is similar to, yet differs in an important respect from several others that have appeared in the literature. In an article to be published in *Urban Studies* (Dunn, forthcoming), I suggest that the degree of entitivity is related to the richness of functional linkages endogenous to the system relative to linkages that the system and subsystems establish collectively with activities external to the system.

Simon (1962) expresses a similar idea. Note this observation: "We have seen that hierarchies have the property of near-decomposability. Intra-component [read intra-system] linkages are generally stronger than inter-component [read inter-system] linkages. This fact has the effect of separating the high frequency dynamics of a hierarchy—involving the internal structure of the components—from the low frequency dynamics—involving interaction among components."

Campbell (1958) also expresses a similar idea, but in different terms. Our discussion contains some of the elements of what he refers to as the "common fate" or "pregnance" of social groups.

The trouble with these statements is that they do not adequately take into account the reality of overlapping boundary conditions. Under these definitions it would be quite possible for a simple system (i.e., one with a limited number of subsystems and subsystem linkages), forming a formal hierarchy under the complete control of the system boundary condition, to display a very high level of exogenous linkage rela-

197

One might go further and point out that in the cases where two or more boundary conditions share the control of a common subsystem, the degree of entitivity is further affected by the degree to which the control objectives are compatible. Of course, if they were so incompatible as to be mutually exclusive, there would be no sharing of common subsystems and no overlapping boundaries. However, it is not uncommon that the advantages of sharing (often associated with the exploitation of external economies) are partially offset by partially conflicting objectives. One system may be dominant in the control it exercises over a common subsystem.

For example, in economics we are familiar with the situation where a joint system serves two or more higher order processes (e.g., the oil-producing enterprise may serve two or more refinery complexes). It is not uncommon for one of the higher order processes to be dominant in the sense that it can exert disproportionate influence over the behavior of the shared component (e.g., more or less determine prices and quantity shares).

In short, the degree of entitivity is also influenced by the extent to which the subsystem control is genuinely shared or is modified by the dominance of competing system controls.

At the same time, the above discussion suggests that the degree of entitivity may also be related to the degree to which the functional overlap may be matched by overlapping controls that can serve to coordinate the overlapping boundaries bringing consistency in the goals and controls of the overlapping systems. This brings us to a point of our story that must be deferred until later in the chapter (pp. 202 ff.).

---

tive to endogenous linkage (e.g., some mineral or extractive process linked to a number of users). According to the concepts cited above the degree of entitivity would be low, but according to the definition in this text it might be very high.

The concept presented in the text appears to be more valid and useful. The statements cited do not define the degree of entitivity in terms of the degree of overlapping boundaries because they do not take as their point of departure the definition of the system in terms of its boundary condition. They tend to think of social systems exclusively in terms of the structure of the activity networks that link components rather than viewing the social system as a control system that identifies the entity of the system with the design and purpose of its control.

MANAGED SYSTEMS VERSUS ECOSYSTEMS:
SUBSYSTEM INDEPENDENCE AND THE DEGREE OF ENTITIVITY

Variations in the degree of entitivity of social systems seem to be associated with another organizational distinction: social systems may either be directly managed or they may function as ecosystems.

In the biological realm organisms are examples of managed systems. Component organs are under the direct control of the volitional and autonomic nervous systems and certain allied chemical and genetic control systems. The behavior of the organ is under the direct control of the boundary conditions of the organism. The effect of activities exogenous to the organism upon the system organs is mediated through their effect upon the organism as an entity with a high degree of closure on control.

In the social realm there are social systems with a sufficiently high degree of functional closure that they can be treated as virtually managed systems. For example, production systems in industry based upon a well-established and stable technology have a tendency to become institutionalized as managed systems. The man, machine, and material components of the system are coordinated and function under an essentially mechanistic plan according to directives issued by management authority. If there is a hierarchy of controls, it tends to be a formal, nested hierarchy. Exogenous changes (e.g., changes in the volume of sales or the cost of components) tend to modify subsystem behavior through their effect upon management directives rather than through direct controls upon component behavior. Such systems tend to be characterized by a high degree of functional closure, or entitivity. The overlapping of system boundaries is minimal and the ratio of the exogenous linkages mediated through the boundary to the exogenous linkages with direct subsystem access is very high.

In the biological realm we recognize ecological systems as well. These are systems whose components are not subjected to direct or formal management control. The systemic behavior arises out of the coadaptations of multiple species that constitute a living interdependent complex. The elements of a set of biological species (plant and animal) define for each other the major aspects of their operating environment. They form the feeding elements and the protective, sup-

199

portive, or reproductive elements necessary for self-maintenance. Any change in either the scale or the behavior of one requires changes in the scale or behavior of others.

These species components or subsystems of the ecosystem often overlap in complex patterns. At the same time they tend to form a system of the whole for several reasons. (1) At some level of complex combination a significant degree of functional closure comes into being. The level and variety of internal linkages is significantly higher than the remaining linkages to exogenous systems (like the cosmological environment or other species whose environmental niches do not overlap). There is established an implicit quasi-boundary where the functional linkages become sparse. (2) The coadaptations of the components of the ecosystem tend to be equilibrating or deviation counteracting. There is an implicit control boundary formed by the deviation-counteracting nature of the coadaptations. (3) There is an implicit joint purpose or objective of the ecosystem. The self-maintenance or survival of the component species is linked to the stability and survival of the ecosystem that forms its immediate operating environment.

Many social systems are ecological in character. The classical general equilibrium market mechanism of economics is, in effect, the description of the function and structure of a social ecosystem. The components of the ecosystem (enterprises and households) are decision units in their own right and are not subject to direct or formal management control. These systems are linked in an interdependent input-output relationship so that each enterprise or household lives in an operating environment formed by the rest. They coadapt.

They tend to form a system of the whole because: (1) at some level of complex combination a significant degree of functional closure is evident; (2) the coadaptation tends to be equilibrating; and (3) there is an implicit objective of maximizing the material welfare of the system components. These attributes establish the boundary conditions of the system.

The political system in a constitutional democracy also tends to be ecological in character.

In truth, however, it is not likely that any social system exists that is an example of a pure managed system or pure ecosystem. In reality most social systems would be more properly characterized as managed ecosystems. This stems from the conscious, purposive nature of human behavior.

For example, it is quite possible for the conventional price and quantity coadaptations of enterprises and households in the market ecosystem to create a situation where aggregate demand and supply are not in balance. Under these conditions the coadaptations can be unstabilizing or deviation amplifying. Runaway inflation or a spiraling recession can result.

In the biological realm the appearance of unstabilizing coadaptations in an ecosystem tends to lead to two consequences—the ecosystem fails to survive or it is condemned to a periodic fluctuation if some force acts to terminate the deviation-amplifying behavior at the extremes.

In the social realm this situation can and has been met by superimposing upon the ecosystem monetary and fiscal management designed to maintain a balance of aggregate supply and demand. The component systems still make their coadaptations, but subject to the constraints imposed by this management. The stability conditions that help define the boundary condition of the system are imposed by explicit and conscious management of the system.

In short, social ecosystems tend to become overlaid with managed superstructures. Deviation-counteracting controls become explicit and purposive rather than the result of a nondirected stochastic process.

Similarly, social systems that tend to become managed systems (like technological production systems) are rarely totally managed. The orders and directives handed down by the management commonly take into account the fact that the human components can interact intelligently with other system components. General directives are often sufficient to support the goals of the system. Human subsystems are permitted to be coadaptive within ranges of choice limited by general directives. Thus, even these systems take on the character of managed ecosystems.

To sum up, social systems differ in the extent to which the control exercised by the boundary condition is proscriptive or contextual in nature. The usual situation is one in which these two forms of control are mixed with a great deal of variability both possible and manifest between systems. This suggests that the degree of social system entitivity is also associated with the degree to which subsystem components are free to act independently of system control. The greater the role of direct subsystem management as proscriptive control relative to contextual control, the higher the degree of entitivity (assuming other factors that affect social system entity remain the same).

COERCIVE VERSUS VOLUNTARY COMPONENT PARTICIPATION:
CONSENSUS AND THE DEGREE OF ENTITIVITY

Another set of factors exerts an influence on the entitivity of the system. Social system components are not like machine system components unless they are, in fact, component machine systems. The human components of social systems cannot all be directed and controlled in the same fashion. Social systems are confronted with a special kind of problem: How can human components be induced to act in ways that are consistent with the boundary conditions of the system?

There are three answers to this question. The first applies when the boundary conditions or goals of the system are congruent with the goals of the human subsystems. Then the individual agrees to subordinate his independence of action in order to share in the behavioral amplification that often attends joint action (or at least to share in the environmental stability that it offers). If the system is a managed system he acts as if he were a machine component responding to proscriptive signals. If the system of control is contextual he voluntarily forms the behavioral response consistent with the intent of the control. This amounts to saying that, in the hierarchy of his goal structure, social system participation takes precedence over lesser order individual goals in a way that does not do violence to the individual's order of values.

The second answer applies to a state in which goal congruence does not occur. Then the individual participation is assured through coercion. This may be physical but, more often, the coercion takes the form of a scale of psychological or economic costs and penalties for nonconformance. Under such circumstances the individual may act as a social system component even though his participation is not a reflection of his perceived goals.

The third answer has a different approach: it is to engineer human goals and motivations in a way designed to make them consistent with the goals of the system—through the exercise of propaganda and mass motivation techniques.

In reality, human participation as social system components is commonly assured by a mixture of these motives. A social system is a more integrated, viable, stable social entity where human participation reflects a considerable consensus. Too high an admixture of coercive or propaganda controls probably reflects the reification of social system goals discussed in the previous chapter. Such systems may appear to function smoothly over short spans and may exhibit an apparent participatory consensus, but it is of a superficial functional order that does not reflect genuine unity of purpose. Such systems have repeatedly demonstrated themselves to be dynamically unstable over longer periods. We will return to this point when we discuss the relationship between social learning and social organization.

SUMMARY

This section has set forth the following proposition: The entitivity of a social system, like that of all behavioral systems, is identified with its boundary conditions. These boundaries are not physical membranes but control processes that organize component activities to fulfill system purposes or objectives. The identification of these entities is complicated by the fact that social systems may vary in the degree of entitivity or coherent control of subsystem activities. Three sets of factors influence the variability of control: (1) the extent to which the control of subsystem components is shared with other exogenous systems; (2) the extent to which controls are contextual and leave to the subsystem options for independent behavior; and (3) the extent to which human subsystem participation is voluntary. Social systems may vary widely in the way in which these three variables become manifest in social system boundaries and social organization.

The Social System as an Activity Network

The concept of the social system as a control system can be complemented by viewing it also as an activity network. The activity network is that which the system boundaries organize and control. To paraphrase the Polanyi statement quoted earlier in this chapter, the social system as an entity works under the control of two distinct principles. The higher one is the principle of the social system's design or purpose; this harnesses the lower one, which consists of the transformation and transfer processes upon which the social system relies. Under the control of the system boundary the component transformation and transfer activities become linked into an activity network designed to serve the goals of the system. The activity network is the behavioral manifestation of the system goals and controls.

NETWORK PATTERNS

The organizational forms that these networks can assume are varied indeed. Elaborate descriptions and classifications would be misplaced in the context of this book but, at the risk of oversimplification, a limited explanation of the variation in network patterns may be useful here.

The bundling of activity systems into linked networks can come about in two ways. The first is when two or more component activities or subsystems representing transformation and transfer processes are joined under a common control for the realization of a common objective. This is the classical hierarchical control system, which is admirably illustrated by physical production systems in modern economics. The hierarchical systems involved in steel making, described in the previous section, form a good example.

The integration of these subsystems in a joint steel-making process may be brought about under the direct management control of a single control center. Alternatively, a series of separately managed components may be brought into consistent organization through the operation of contextual controls such as market bids, etc., or through a series of mutually agreed upon contracts. Thus, the institutional forms that monitor the linkages and activity flows vary according to the degree of subsystem independence and consensus, but the result is a set of activities that form a classical control hierarchy. These may form a "narrow-span hierarchy" (Simon's term) if a series of heterogeneous activities form a vertical sequence of input and output linkages (e.g., activities that produce coke, pig iron, steel and blooms and billets in that order). They may also form a "broad-span hierarchy" where a series of heterogeneous or homogeneous activities form parallel linkages with a common outlet activity (e.g., coal, iron ore, and limestone in the production of pig iron; or a set of independent milk producers into an urban milkshed). Where these broad-span hierarchies are not under direct management control they often join into "producers associations" or other forms of joint action so that, in forming their linkages with emergent control levels in the supersystem hierarchy, they can act as a single subsystem rather than a set of subsystems.

The second way in which activity systems can become bundled into a linked network is manifest where two or more control systems share a common component activity or subsystem. This creates the situation where a subsystem falls under the partial control of two or more boundary conditions. The economist knows that this component sharing is a result of economies of scale and specialization and, occasionally, monopoly subsystem control of a scarce resource. The component activity thus shared may be either a transformation activity or a transfer activity. Each has led to a distinct kind of network configuration that is typical of modern societies.

Where the shared subsystem component is a transformation activity, one result is the industrial complex familiar to us all today. An example is the metals industry complex where a whole set of fabricating systems shares a common source of semifinished steel or aluminum. Another example is the petrochemical complex where an extended set of chemical producers (drugs, insecticides, paints, etc.) share a few common feed stocks.

Where the shared subsystem component is a transfer activity or channel, one result is the urban complex, the principal function of which is to provide for common or shared access.

System complexes that are formed through component sharing are quite different from classical control systems.[4] They tend to be low entity agglomerations or ecosystems where each participant in the sharing forms an operational environment for the other. Variations in the behavior of one can affect the conditions of the sharing and have effects upon the other. They are bound into a linked system by the implicit goal of realizing the economies of shared specializations. A weak level of general equilibrium control tends to develop that is mediated through the shared components.

At the same time the emergence of such linked systems in social evolution very often has led to their further transformation into more explicit control systems. In an industrial complex the cartelization of the systems linked through joint demand or supply relationship is not uncommon. More important, emerging urban complexes that serve primarily to form the efficient transfer links between systems have yielded a whole series of superimposed governmental and market systems that establish and enforce common rules to assure the efficiency of transfer and the viability of system organization.

Thus, social organization as we know it is a complex fabric of mixed networks. A variety of high entity control systems is linked together into low entity ecosystems through component sharing. This generates both spatial and functional agglomeration. These tend, in turn, to become overlaid by medium entity control systems that supply and enforce contextual rules that assure that the ecosystem is stable and can function efficiently.

PHYSICAL VERSUS SYMBOLIC NETWORKS

Before we leave this view of the social system as an activity network, an additional distinction should be made. The networks that represent the morphology of social systems may take the form of either physical or symbolic networks. Until recently most of the attention in the social sciences has been focused on the organization of physical

---

[4] In this connection it is worth noting that the concept of the urban hierarchy in central place theory is not a classical control hierarchy. It grows out of variations in the scale economies of shared transfers and other final demand servicing functions. Larger urban areas can, thus, enjoy a broader range of shared functions than smaller ones.

networks. This has been particularly true of the economist who has been primarily concerned with organization and management of the thermodynamic throughput of society. But the very concept of control is a symbolic one. A system boundary is a behavioral design. Its implementation requires the generation and transmission (we are still dealing with transformations and transfers) of control signals that govern the behavior of the physical components. Thus, the familiar network of physical transformations and transfers is matched by a dual network of symbolic generation (management) and transmission (communication). Most familiar control systems are a composite of physical and symbolic networks. For example, the flow of goods in the industrial system is matched by a counterflow of money and orders which are symbolic signals. It is worth noting that the physical and symbolic networks are not typically congruent. Physical and symbolic transfers commonly utilize different modes of transfer and follow different channels.

The topology of the symbolic networks that form this dual network is typically quite different from that of the physical system. The line costs or channel costs in transferring symbols are different from those involved in physical transfers. Management (the production of symbols) experiences different scale economies from those associated with the production of goods. These differences can be important in explaining the differences in network topology for mixed network systems like urban systems. This is a neglected aspect of study that is well worth additional attention.

Social System Maintenance versus System Transformation

It has seemed important to identify the nature of a social system as an entity and the way the combination of high entity and low entity social systems generates a convoluted array of activity networks, because it is a topic to which little attention has been paid. Moreover, these concepts are an essential base for carrying out the main purpose of this chapter.

It is now time to remind ourselves of a point made in chapter III. Social behavior does not always lead to social learning. Historically, the major portion of behavior-analyzing behavior and behavior-organizing behavior has been directed to social maintenance. It is homeo-static in character. Once a social system is formed, most of the action that takes place under control of its boundary is directed to carrying out the goals and implementing the behavioral design that defines the system as an entity.

The social entity may behave much of the time in the fashion of the machine system open to resource exchange described in chapter I, or like a lower order animal organism that reacts to its operational environment in the process of seeking out the thermodynamic through-put that maintains it. Such a behavioral system may alter its manifest behavior as it encounters environmental changes at its input or output termini, but it does so within the context of an established behavioral mode and system objective.

Thus, social behavior may be absorbed with (1) the task of collecting and examining the evidence that reveals the states of the operating environment at input-output termini and that maintains the organization of endogenous component behaviors, (2) interpreting the meaning of these signals in the light of system objectives, and (3) managing the component behaviors so as to reconcile the system objectives in the most efficient way with the operational environment. Such social behavior is absorbed in the service of maintaining a steady state system—a system whose throughput cycle operates continuously to carry out the prescribed input-output mode.

But all social behavior is not reactive, and absorbed with maintaining behavior. The human cognitive process has the capacity to be active and anticipatory in nature, and to promote its goals through behavioral innovations. It has the capacity to discover new knowledge and new modes of behavior through play—that is, through behavior that seeks understanding and experience not immediately relevant to self-maintenance. It also has the capacity to modify old goals and develop new. These creative processes are manifest in the behavior of social systems, as well. Social systems sometimes encounter changes in their operational environments that create major discrepancies between their objectives and behavior—discrepancies so large that routine social analysis and social management cannot satisfy system goals by reorganizing traditional behavioral modes. Failure to satisfy the goals of the system constitutes a problem that stimulates the creative attributes of behavior-analyzing and -organizing behavior. New kinds of evidence are sought about the behavior of both external and internal environments. This evidence is examined for clues to the nature of the problem and to modifications in behavioral mode that may better satisfy the goals of the system. *Behavior-organizing behavior may be pressed into the service of designing and implementing new behavioral modes.*

As we have seen, this problem-solving, opportunity-seeking attribute of human social behavior gives rise to behavior directed to changing behavior. This is not a separate and independent characteristic of social behavior. Behavior directed to both analyzing and organizing behavior must be pressed into the service of behavior-changing behavior.

We are faced with the paradox that social organization can be addressed not only to maintaining but also to transforming established behavior. However, in our discussion of the boundaries and networks that constitute social organization, through both language and example attention has been focused on organizations engaged in the maintenance of social throughput. This is the habit of virtually all social science treatment of social organization. Now we must focus on the relationship of social organization to social learning or social change. Here there is a double focus. We need to consider how the process of social learning acts to transform social organization; we also need to ask whether there is something different about the social organization that organizes and directs processes of social learning. In the remaining sections we turn our attention to these issues.

The Effect of Social Learning upon Social Organization

Consider, first, the way in which the ongoing process of social learning transforms existing social organization. Since, in the previous section, the social system has been viewed as both an entity and an activity network, here the organizational change attending social learning will be represented as both a process of entity redefinition and network transformation.

ENTITY REDEFINITION

The point of emphasis here is the fact that social reorganization is an essential element of evolutionary experimentation; we are simply reexamining the processes discussed in chapter IV from another perspective. Accordingly, we would expect the redefinition of social entity that attends normal problem solving to be different from that associated with a paradigm shift. Two reminders are in order before we examine the relationship more closely.

First, we need to remind ourselves that social change does not necessarily involve social learning at the group or system level. Homeostatic adjustments quite often do not necessitate changes in social system objectives or controls or any novel forms of component behavior. Social systems are commonly organized in a way that permits them to maintain their boundaries and controls within some limit of variation with respect to their operating environments. However, these adaptations to parameter changes are predetermined by the design of the system and the instrumental goals and performance criteria it embodies. They are essentially mechanical adjustments brought about by deviations between the readings of an environmental sensing unit and a pre-set performance standard that set in motion control signals yielding homeostatic changes in some dimensions of system performance.

The changes usually take the form of network flow adjustments. These may be changes in the levels and the mix of the flows. They may imply changes in the pattern of the functional links formed by subsystems, even to the point of yielding changes in the pattern of network channels. However, these adjustments do not involve social system learning.

At the same time, it is possible that social learning is involved in a limited way at subsystem levels. Changes in network flows that carry a system component beyond some capacity limitation might require capacity changes—i.e., net capital formation. This might require the individual learning necessary to augment the existing stocks of human and physical capital. It would not necessarily require social system learning that introduced novel behavior or modified the basic design of the control system.

Second, we need to remind ourselves that social learning does not necessarily generate social system learning or social reorganization. Men, social organizations devoted to science, and even social system control centers may learn in the sense that some units conduct analysis and acquire new knowledge. However, until such knowledge becomes embodied in a developmental hypothesis, incorporated into control system design and manifest in experimental social controls organizing component activities, social reorganization or social system learning does not take place. Learning has to be addressed to problem solving through social action before social system learning takes place.

Normal Problem Solving: Redefining Subsystem Boundaries. There are times when environmental changes cannot be effectively countered with homeostatic adjustments. System goals or standards can no longer be satisfied by the operation of existing controls. Periodically, new technology may appear that offers new options and gives rise to a shift in system objectives to embrace the new option. Whether the source of change originates with changes in the environmental parameters or the behavior options arising out of new knowledge, a discrepancy between system goals and system behavior occurs that requires some innovation in the modes of behavior and their control.

In many instances systems can effect this kind of change by modifying subsystem behavior. This is the classical form of the incremental, normal problem solving discussed earlier. The reorganization of component behaviors to include novel (to the system) forms and controls is carried out without any major change in the total system objectives and controls. Change is brought about through component tinkering.

From the point of view of the activity component or subsystem involved, the change may be total and dramatic. The subsystem boundary, the components included in its system design, the network that links them, and the controls that order their behavior may be drastically modified. However, even from the point of view of the subsystem, it would not be proper to characterize this as a paradigm shift because *the reorganization of the subsystem is brought about under external control.* The motivation, the context, and the direction for change are exogenously provided by the total system boundaries. The change is either directly prescribed and managed by the total system controls, or the quasi-independent decision units at the subsystem level guide it in accordance with a favorable stimulus provided by the total system context.

This kind of normal problem solving is relatively low risk in character. Although the novel modes of behavior and controls may be new

211

to the system, they have often been pretested in exogenous systems. It can be pretty well anticipated that their introduction will narrow the gap between total system goals and performance. Even where the subsystem reorganization is more experimental in form the evolutionary experimentation can be carried out in a manner subject to close monitoring and in-process modification by the total system controls. The experimental modifications are subjected to careful and continuous reality testing in terms of the desired total system goal convergence.

It is important to realize that these innovations may have a broader scope than commonly emphasized. Greatest emphasis in normal problem solving has been placed upon the role of technology narrowly conceived—i.e., the innovation of novel physico-chemical subsystems. The conventional image of normal problem solving is the replacement of an obsolete man-machine subsystem with one that is more efficient or the grafting on of a new technological capacity.

To give proper balance to the reality of normal problem solving it should be remembered that more can be involved than redesigning, reorganizing, or replacing activity components. It may take the form of modifying the nature of subsystem control. This can often be done without substantially modifying the total system boundary. The mix of proscriptive and contextual controls at the subsystem level and the nature and scope of the decisions for which a human component consensus is sought can be altered. Such changes in the forms of control and the areas of activity in which they are operative may have an important effect upon the contribution that subsystem reorganization can make to total system goal convergence. The technology of social control and consensus formation is still in such an imperfect state that characteristically its tremendous importance for social system problem solving is either overlooked or de-emphasized. It is clear that social science must expand its interest in this domain.

Paradigm Shifts: Redefining Total System Boundaries. Situations can occur in which normal problem solving is not enough. Subsystem reorganization is not adequate to cope with the problem or, paradoxically, subsystem reorganization comes to constitute the problem. In the latter case the system must turn to a redefinition of its own boundaries.

Its goals, and the design of the controls that harness component activities to its satisfaction, must be altered.

Once this is accomplished the system may return once again to homeostasis and normal problem solving, but it is a distinct shift that is different in character from either. It differs from normal problem solving because the change is something that it does to itself rather than something that it does to something else (e.g., a subsystem) or something that is done to it (e.g., by a supersystem). It is a recreation of its "self-image," a remaking of its own boundaries. It is an alteration in the purposes and operational context of the whole system. It is commonly an evolutionary experimentation that is undertaken without the benefit of a supersystem context that serves to guide the transformation and provide for the systematic reality testing of it. The reality testing has to be carried out in terms of the extent to which the paradigm shift or boundary shift creates behavioral designs that prove adequate in the service of new goals.

Historically, these boundary shifts in total systems seem to have come about under two different circumstances. Under one of these the paradigm shift or system reorganization emerges out of normal problem solving activity. A social system may undertake a series of subsystem reorganizations designed to harness new technology to the satisfaction of system goals or designed to extend the system's adaptive range in the face of environmental change. The cumulative effect of these, however, may bring the system to a threshold where the modified subsystem behavior calls into question the operating context which fostered these changes in the service of its own objectives. The total system boundary is challenged. The objectives and controls of the system undergo a metamorphosis.

Take an example from the field of economic enterprise. A sheet metal fabricating system may visualize its role as a supply contractor for the construction industry. As such it may do sheet metal roofing, produce air conditioning ducts, gutters, etc. In the interest of meeting its profit and efficiency goals it may introduce into a fabricating division new metal forming machines. This may proceed to the point where, with minor changes, it could enter a wider or different field of metal fabricating. By modifying its own image, building a new set of external linkages, and modifying its internal controls it could generate

213

a major shift in its operating context. It might shift to become a supplier of metal chassis for electronic assemblies or supply final demand directly with such items as mail boxes or outdoor grills.

The second historical circumstance leading to paradigm shifts occurs when the system is confronted with an environmental change so great that to modify subsystem behavior through the exercise of established total system controls seems completely inadequate to assure system survival. To select another economic enterprise example, this is what happened to the carriage producer when confronted with the development of the automobile. Those that survived did so by making a radical shift into automobile production or some other enterprise activity. This required a complete shift in the operating context and the organization and control of subsystem activity.

The Principal Problem of Social Organization for Advanced Societies. This discussion brings us to the point where we can more clearly identify the principal problem of social organization that confronts advanced societies. It can best be seen by contrasting the organizational consequences of normal problem solving and paradigm shifts. In normal problem solving, system reorganization is purposive, anticipatory, and controlled.[5] In contrast, paradigm shifts have historically been primarily reactive, unanticipated, and uncontrolled. They tend to arise out of a reaction to exogenously and endogenously generated boundary crises. They are frequently unanticipated by the management élite of the system. The reorganization is often defensive in character—that is, it is directed toward preserving and extending the life of the system in the face of change rather than taking the form of a directed self-transformation in pursuit of some higher order goal.

We appear to have arrived at a point in social history where this kind of uncontrolled social reorganization is taking on a rate and form that seem to threaten the viability of the social process itself. The reason for the special nature of the threat in advanced societies is the fact

[5] Social systems may vary substantially in their capacity to carry out normal problem solving, but they rely essentially upon subsystem reorganization. Social systems with extremely low degrees of entitivity may not have a sufficiently centralized control to undertake controlled endogenous reorganization. However, normally some degree of either contextual or prescriptive control is exercised over the reorganization.

that we have formalized and mastered the process of normal problem solving in the physical sciences and in the design of physical systems to the point that this very process is generating an accelerated and escalating series of boundary crises for established social systems.

The problem solving that goes on under the accepted theoretical paradigms of classical science is abstracted from its consequences for social reorganization. Even where physical science experiences shifts in its theoretical paradigms "paradigm agreement looks like a disinterested matter of option for an objectively compelling theory. It does not look like an active social historical problem." (Becker, 1968, p. 362.) More important, even at the level where the knowledge of physical system relationships thus gained is incorporated into system design through classical engineering, it is the behavioral options of machine systems that are the focal point of conscious innovation. It is the machine system components of social systems that are the initial objects of redefinition and reorganization. No immediate and direct challenge to established social system boundaries is understood or contemplated. All is carried out in the objective pursuit of "efficiency" in the realization of established objectives.

It should be apparent, however, that the behavioral amplification that attends the use of modern machine system artifacts as components in social systems requires that they be embodied in man-machine systems. Their redesign in the interest of normal problem solving tends to generate a cumulating sequence that yields a threshold where the adequacy of the total system boundary is challenged. An organizational crisis arises out of changes in the system's endogenous environment that were pursued in the interest of efficiency. With the accelerated rate of advance in science and engineering, the social systems that organize human activity are being presented with a rapidly increasing and recurring set of organizational crises requiring boundary shifts for their resolution.

Furthermore, when the inventiveness of science arrives at a point where, within a short span of history, we are confronted with such radical innovations as the automobile, the airplane, the rocket, the computer, atomic energy, artificial birth control, and when we are on the threshold of human organic and genetic engineering, we are faced with behavioral options that offer a direct frontal challenge to established

215

social organizations and goals. Their application to social life requires major shifts in social organization. Their impact surpasses anything that can be encompassed by normal problem solving. The symbolic environment of social life is changing so rapidly and radically that we are faced with the need for a continuous series of paradigm shifts. This frontal challenge is only now being dimly perceived. It was ignored for so long because of our habit of viewing science and technology as objective pursuits removed from the value-laden decisions of the everyday social process.

Radical and frequent shifts in the boundary conditions of social organizations have been created by social learning in the domain of traditional science and engineering. But recent experience suggests that the rate and radical nature of these transformations are threatening social stability. There is also the threat that the burgeoning knowledge and the power of the human artifacts that it yields might, through the reification of the social system, be applied to nonhumanistic goals.

All of this poses the big issue of the modern age: Can these emergent aspects of social change be made more purposive and subject to social control? Can the boundary reorganizations or paradigm shifts be set in motion by the system itself—not as a reaction to radical environmental change, not just because the old goals and controls have become unstable and irrelevant, but as an active exercise in building a new system capable of fulfilling human goals better?

The classical science answer would be No. To attempt this level of social control would imply that scientific pursuits and social change be teleological in character when we know we cannot anticipate the future states of a nondeterministic system such as the social process. Any attempt to do so would simply distort the process by attempting to force it into a predetermined mold. It would also have the effect of threatening the creative freedom of science and the social processes generally.

Chapters IV and V suggest that this need not be the case. Social development can be anticipated, purposive, and subject to reality testing and still be experimental and creative in nature. The whole issue focuses upon those instances of paradigm shift or total system reorganization. When these shifts are reactive in character, as has been common, the transformation can be erratic and traumatic. It is reasonable to suppose that if we understood the nature of this process adequately we could devise organization and controls that would make the transformation more efficient, once the need for reorganization was established. This raises the issue of the role of social organization in social learning, a topic that will be treated in a following section.

More important is the issue of the purpose of control. One cannot derive much assurance from the possibility of developing social procedures that can help social reorganization follow more efficiently the implications of new behavioral options. Some of those options could lead to our destruction. The issue is: Can total system reorganization be provided with a higher order paradigm that guides the choice among emergent options in a way similar to the control exercised by system boundaries over subsystem reorganization in normal problem solving? It was suggested in the last chapter that such an over-arching paradigm or social value does exist in the need to make social organization serve human needs. Paramount among them is the need to serve the human growth motive—to assure that social change is consistent with the requirements of human development revealed by a human science.

This constitutes a distinct shift in our orientation to social problem solving and social organization. We have been accustomed to thinking in terms of finding processes consistent with states—i.e., reordering activities to serve established goals and controls more efficiently. The paramount problem of our time confronts us with the necessity of finding a state consistent with a process—i.e., developing forms of social organization and a humanistic ethic based upon a human science that can be used to organize the process of social learning in the service of the process of human development.

NETWORK TRANSFORMATION

The effect of social learning upon social organization can be discussed from the point of view of activity networks as well as that of entity reorganization. As we have seen, social organizations as control systems are made up of a complex network of functionally linked activities. It is this network that the system boundaries organize and control. A variety of high entity control systems with their internal functional networks are linked together into low entity ecosystems through component sharing. These networks have both a functional and spatial dimension. Any change in social organization and control is reflected in changes apparent in the structure of the network patterns. Let us, then, consider the way in which social learning can modify the functional and spatial network patterns.

It is useful to note in this connection that any change in social behavior generated by social learning is effected through a change in either a transformation process, a transfer process, or both. There are no exceptions. This is just as true of symbolic components of social learning as it is of the processes of transformation and transfer associated with material and energy transformations. Changes in the technical nature of these transformations and transfers wrought by social learning modify the form of the links and the structure of the networks. It is this network reorganization that has captured more of the attention of social science than the nature of social boundaries and controls that represent the opposite face of the coin.

For example, when, during the course of development of the steel industry, transformation processes (steel converters) were changed to allow for the use of scrap steel as input, a new set of transformation processes emerged, bundled together by the shared objective of producing steel. The scrap dealers in urban centers who sorted, graded, consolidated, and baled scrap became an important linked process. At the same time the transfers internal to this complex of transformations were substantially modified, the levels of transfer with earlier established sources of ore, etc., were modified by substitution relationships, and new transfer relationships (for assembling scrap) were generated. Where these changes in levels and patterns of transfer involved the use of general purpose transfer channels they tended to modify the capacities of existing channels and, perhaps, cross thresholds of scale economies giving rise to new channels. Where changes in the patterns of transfer involved special purpose transfers internal to the production complex, new channels would of necessity be established.

It sometimes happens in a case like this that the new set of transfer relationships implicit in a change in transformation processes may in time modify the location of some of the basic transformations and, thus, radically alter the patterns of transfers linking the transformations in the complex. A radical reorientation of this sort took place in this illustrative case. The role of scrap in the complex had the effect of strengthening the pull of market centers for steel and generated several major shifts in the production centers as industry capacity expanded.

In such a case the change in a transformation process may have no direct or indirect effect upon the technology characteristic of any transfer processes, but may have a substantial effect upon their pattern and the ways in which they are shared or bundled. This may have a feedback effect upon the location of the transformation process itself. Conversely, a change in a transfer process, in addition to reorganizing multiple purpose transfer channels, may change the orientation and bundling of transformation processes without having any effect upon the technological characteristics of the transformations.

For example, the development of rail transportation as a supplement to water carriers for heavy and bulky material transfers resulted in the generation of a whole new network of general purpose transfer channels. This, in turn, opened up sources of material resource inputs, potential production sites, and new market relationships, which modified the set of transformation processes that were bundled into a common transformation complex. It even fostered the creation of entirely new complexes. Similarly, the invention of the automobile has created whole new networks of transfer. It has increased the tendency of central-place or final-demand servicing network circuits to centralize into larger urban centers. At the same time it has enabled some manufacturing systems to find satisfactory transfer services in small urban centers.

It is possible for a change in a transformation process to have either a direct or indirect effect upon a transfer process, and vice versa. Direct effects come about because transfers involve the use of material and energy converters in the service of the movements of people, artifacts, and ideas. An improvement in an energy converter inspired by the solution of a transformation problem in industry occasionally may be applicable to a technological modification of transfer processes. The development of the steam engine was one such process generic to both types of use.

The indirect effects are likely to be even more important. Each change in a transformation function may give rise to a new or modified transfer pattern, and thus set the stage for transfer innovation in its solution. For example, the development of liquid petroleum as an energy source created novel transfer problems that led to the development of specialized carriers (e.g., tank cars) and even to the development of an entirely new transfer technology and a new transfer network (e.g., pipelines). Similarly, the development of a new transportation mode and its resulting networks (e.g., the railway) generated such an important modification in the scale of the demand for steel that it created new production problems and opportunities leading to technological changes in the transformation processes. In short, each change in a transfer or transformation process creates new functional relationships and new linkage patterns that often raise newly perceived transformation and transfer problems for solution by social control processes.

These examples are taken from the traditional networks formed by material and energy transformations and transfers, but the same kind of phenomenon can be traced in the networks that form the symbolic networks and the organized components of social learning. Members of the research community, for example, are acutely aware of the fundamental changes in the structure of research systems caused by the impact of the computer. The effect is to redefine the way in which research activities are joined by common objectives as well as the patterns of transfer and communication that link them. Indeed, the computer is an instance of an innovation that affects both transformation and transfer processes directly through its effect upon both research methods and the communication of research findings.

In short, whatever the class or structural type of functional network at issue, the changes in transformation and transfer processes generated by social learning modify the way in which transformation processes share common objectives and common transformation and transfer components. The functional and spatial arrangements become modified. Activities shift in functional relationship causing old networks and agglomerations to break up and new ones to be established.

The Effect of Social Organization upon Social Learning

In the previous section we considered the ways that social learning affects social organization. We saw that activities addressed to social maintenance, normal problem solving, and paradigm shifts all affect the social system as control systems in different ways and at different levels. We saw how the functional and spatial characteristics of activity networks become modified. Now we want to reverse our orientation and inquire into the ways in which social organization affects social learning. Two aspects of this view of the social process are worth emphasis.

ORGANIZATIONAL ANOMALIES AND EVOLUTIONARY HYPOTHESES

The first aspect can be disposed of quite briefly, but it is an important point to make. All social problems and all social problem solving emerge out of the anomalies of social organization. Social system boundaries are often found to be ineffective activity organizers. The controls are not adequate to serve the goals. This discrepancy is a constant source of frustration and dissatisfaction and leads to efforts to improve the relationship by means of evolutionary hypotheses that are tested in action. These evolutionary hypotheses have to do with the modification of relationships between system goals and controls. So we see that social system learning emerges out of the anomalies of social organization and acts through an experiment in social reorganization. It is the organizational boundary that is both the source and object of social system learning.

ORGANIZATION FOR SOCIAL LEARNING

Organization is related to social learning in a second way. Learning activities—those evaluative and reorganizing activities—are themselves subject to organization. It matters a great deal, therefore, how behavior directed to changing behavior is provided with social organization. How do we organize for learning? Does organization for social maintenance differ fundamentally in character from organization for social learning? Is social learning organized in such a way that its goals and controls are consistent? How can behavior directed to changing behavior be organized and provided with goals and controls in a manner that will give to human society greater control over its destiny?

Answers to these questions must surely be of critical importance. But social science has not yet mounted the kind of effort devoted to analysis and evolutionary experimentation necessary to a resolution of the underlying issues. Our treatment here, therefore, must rely on work of an exploratory nature that has been conducted recently.

Likert, Pelz, and Bennis.   The three studies cited here illustrate the emerging concern for these matters in the social sciences, and two of them provide some empirical support for the conceptual generalizations offered in this and preceding sections.

In an analytical essay, Bennis and Slater (1968) call attention to the fact that we live increasingly in a "temporary society." Social systems are temporary in the sense that they undergo frequent modification and reorganization. Under these circumstances they claim that the traditional bureaucratic forms of organization do not work well.

They refer to bureaucracy as "a useful social invention that was perfected during the industrial revolution to organize and direct the activities of a business firm. ... The bureaucratic 'machine model' was developed as a reaction against the personal subjugation, nepotism, cruelty, and the capricious and subjective judgment which passed for managerial practices during the early days of the industrial revolution. Bureaucracy emerged out of the organization's need for order and precision and the worker's demand for impartial treatment. Most students of organization would say that its anatomy consists of the following components:

> A well defined chain of command.
> A system of rules and procedures for dealing with all contingencies related to work activities.
> A division of labor based upon specialization.
> Promotion and selection based upon technical competence.
> Impersonality in labor relations.

It is the pyramid arrangement we see on most organization charts" (pp. 54–55).

Bennis and Slater point out that "Bureaucracy's strength is its capac-

ity to manage efficiently the routine and predictable in human affairs.
... with its well defined chain of command, its rules, and its rigidities
[it] is ill-adapted to the rapid change that environment now demands"
(p. 56). The authors go on to point out that increases in social system
size reach the point where pyramidal organizational structures become
unwieldy. They conclude by asserting the need "not only to humanize
the organization, but to use it as a crucible of personal growth and the
development of self-realization" (p. 59).

The authors advocate a new approach to organization they would
characterize as "organizational revitalization," a complex social process
which involves a deliberate and self-conscious examination of organizational behavior and a collaborative relationship between managers
and scientists to improve performance" (p. 60).

They have this to say about the consequences for organization:
"There will be adaptive, rapidly changing *temporary systems....*
groups will be arranged upon an organic rather than mechanical
model, meaning that they will evolve in response to a problem rather
than to preset, programmed expectations. People will be evaluated
not vertically according to rank and status, but flexibly according to
competence. Organizational charts will consist of project groups rather
than stratified functional groups" (p. 98).

And, "The language of organizational theory in most contemporary
writings reflects the machine metaphor: social engineering, equilibrium, function, resistance, force field, etc. The vocabulary for adaptive
organizations requires an organic metaphor, a description of a process,
not structural arrangements" (p. 120).

The other two works cited here differ from the Bennis-Slater essay
because their generalizations are supported by the results of empirical
research studies. The authors devoted considerable ingenuity to meas-
uring the relationship between organizational forms and productivity
or performance.

The first of these is the Likert (1967) study of the relationship
between business management and productivity. The author is particularly concerned with the management of the human components of an
organization, and identifies four systems of organization ranging between "exploitive authoritative" (System 1) at one extreme and "par-

223

ticipative group" (System 4) at the other. It was observed that "Those firms or plants where System 4 is used show a high productivity, low scrap loss, low costs, favorable attitudes, and excellent labor relations. The converse seems to be the case for companies or departments whose management system is well toward System 1. . . . Shifts toward System 4 are accompanied by long range improvement in productivity, labor relations, costs and earnings. The long range consequence of shifts toward System 1 are unfavorable. . . . System 4 is appreciably more complex than other systems. It *requires greater learning* and appreciably greater skill to use it well, but it yields impressively better results . . ." (p. 46, italics mine).

Likert emphasizes that one of the big shortfalls in traditional business management is the fact that standard techniques for financial and physical asset accounting are not matched by satisfactory techniques for human asset accounting. As a consequence, asset building in relation to human components is not properly recognized and accounted for. This tempts managers in trouble to shift toward System 1 because it yields temporary improvements in traditional accounts stemming from the liquidation of human assets that are not characteristically accounted for. At the same time, it disguises the long-run gains that can be derived from System 4. This emphasizes the great importance of developing a mode of accountability that can reflect the quality and performance of human components.

The third point of interest to us in Likert's study is his conclusion about organization. He writes (pp. 156–58):

Virtually every large company faces, in more or less serious form, the problem of whether to organize on a functional basis or on a product or geographical basis or to try some compromise solution. The requirements of both specialization and low unit costs achieved by large-scale operations press for a functional form of organization. But it is not easy for a large, highly functionalized organization to achieve effective coordination.

Decentralization is becoming . . . an inadequate solution as technologies become more complex and ever more extensive functionalization becomes essential. Decentralization, furthermore, does not eliminate differences among staff or among departments; it merely changes the relationship of who differs with whom about what.

A satisfactory solution requires an organization which can have extensive functionalization and which can also resolve differences and achieve efficient coordination on a product or geographical basis. ... These decision-making and influence processes must be able to achieve coordination in spite of initial and often substantial conflict coming through two or more channels or lines.[6] At least four conditions must be met by an organization if it is to achieve a satisfactory solution to the coordination-functional problem.

(1) It must provide high levels of cooperative behavior between superiors and subordinates and especially among peers. Favorable attitudes and confidence and trust are needed among its members.

(2) It must have the organizational structure and the interaction skills required to solve differences and conflicts and to attain creative solutions.

(3) It must possess the capacity to exert influence and to create motivation and coordination without traditional forms of line authority.

(4) Its decision-making processes and superior-subordinate relationships must be such as to enable a person to perform his job well and without hazard when he has two or more superiors.

Likert expands upon the organization structure alluded to above, concluding that what is needed is a "multiple, overlapping group structure sharing horizontal linkages as well as vertical linkages." Individuals or work groups perform cross-function coordination by playing a role simultaneously in two or more related subsystems. Likert apparently is recommending the establishment of overlapping controls to match the overlapping boundaries of social systems.

The other empirical research study was conducted by Pelz and Andrews (1966). This was not a study of business organization, but a study of "Scientists in Organization." The authors were concerned with "what constitutes a stimulating atmosphere for research and development."

Although conducted in a different domain, this study reinforces the results of the Likert study. The authors state that "Much of the writing about research organizations assumes that freedom and coordina-

---

[6] I.e., it must be able to solve the problem of overlapping boundaries and human component consensus. (*Author's note.*)

225

tion are incompatible. Our data suggest that some combination of both is not only feasible, but helpful for the scientist himself, that is, where he involves several other people in shaping his assignments, but keeps substantial influence over the decision process. . . . Likert's concept of management by overlapping groups is consistent with our results." In short, the authors relate scientific productivity positively with broad span, face-to-face interaction and overlapping activities. This is explained on the grounds that in this way the scientist "can learn the major goals that are important. . . [that is,] the relevance of the scientists' efforts to the solution of important problems is directly apparent. . . . there is the stimulation of having other people interested. . . . They can prove a testing ground [and a] challenge to sharpen ideas." This interacting relationship with people in different roles exposes one to different viewpoints and a wider source of information. Furthermore, "the scientist performed best when he utilized two or three different skills, and faced both scientific and applied problems in his work." Productivity is directly related to "the feeling of intense involvement in one's work." The scientists became "more involved and committed to" the research that they shared in formulating (pp. 32–33, 76, 88).

Organizational Contrasts. This brief review of some of the latest innovative work in the field of organization conveys some idea of the start that has been made toward understanding the effect of organization upon social learning. Against this background let us return briefly to the perspective that relates the entitivity of the social system to its function as a control system.

One thing should now be clear. The formal nested control hierarchy (what Bennis and Slater refer to as bureaucracy) is an efficient control system only where the control problem is restricted to homeostatic adjustments and where the system does not experience any boundary overlapping except where the common or shared component is not subject to incompatible controls. This appears to be the only instance in which the formal chain of command (which passes prescriptive orders down through a series of hierarchical control centers) can perform properly. In any other circumstance, controls of this type cannot adequately satisfy system goals.

226

Take the case where boundary overlap occurs and the system shares a component under partial control of another boundary. If the joint interest of the two boundaries in the subsystem component is not fully compatible, a formal control hierarchy may send a control signal to the shared subsystem designed to serve the homeostatic goals of the system. However, that signal may be countermanded by incompatible control signals from the competing boundary so that homeostasis is not achieved for either system or is achieved by the dominant system alone.

Such a circumstance, however, escalates the management problem to a new level—from the management of homeostasis to problem solving (even if the problem is perceived as the restoration of homeostatic controls). Likert makes clear that the solution to this problem requires modifying the traditional "line of command" controls. Actually, one might visualize solving the problem and preserving the formal centralized control structure by superimposing a higher order control boundary upon the two boundaries competing for subsystem control. This higher order boundary could then prescribe the way in which the competing boundaries will share subsystem behavior in accordance with some higher order goal or criterion. The implication of the Bennis-Slater allusion to size, however, is that where functional overlapping is extreme, the attempt to resolve the problem of incompatible boundaries by extending the logic of the nested hierarchy would soon become unwieldy and collapse. In a complex system this would be a formidable solution. It can be seen to be all the more complex when we recognize that the human individual is a fundamental component and that the individual has his own boundary conditions, which may be inconsistent. The solution yielded by deepening the hierarchical superstructure becomes untenable as a general solution.

The solution, then, lies in the direction outlined by Likert. One cannot always seek to force the coordination of a set of overlapping boundaries by subjecting them to a higher order directive control. Then one must seek to institute overlapping controls in the areas of overlapping function. By their nature these cannot be prescriptive controls. They take the form of negotiated arrangements through a cooperative dialogue that brings the controls affecting the shared subsystem into consonance.

All of the problems social systems face do not find their origin in overlapping boundaries. Changes in both endogenous and exogenous environments may give rise to serious inconsistencies between system goals and system controls. The problem thus exposed may or may not apply to systems with overlapping boundaries, but in either case they require the formulation of developmental hypotheses leading to experimental system reorganization. The structure of organizations needs to allow for some problem-oriented groups that can act to bring about network realignments and the readjustment of goals and controls.

We conclude that the more a social system is confronted with problem-solving situations and the need for social learning, the less it can rely on formal prescriptive hierarchical controls. The learning or reorganization process seems to be better served by organization structures that are characterized by overlapping boundaries and contextual controls (where these are supplemented by procedures designed to produce consensus). Some such organizational pattern seems necessary to provide the freedom for creative adaptability together with the degree of control (including self-control) essential for social system cohesion and viability.

One is prompted to wonder what kind of variations in these patterns can exist. For example, are there significant differences in the organizational patterns relevant for systems engaged in normal problem solving versus paradigm shifts? Are there circumstances and ways in which formal patterns of control and the more flexible controls form a congenial mix? Are there differences in kind and degree between the organization of learning systems dealing with problems of physical system modification and the organization of learning systems dealing with modifications in symbolic systems (e.g., pure versus applied science, system analysis versus system design and implementation)? Are there, perhaps, layers of organizational control where the social system will embrace a control level engaged in homeostatic controls (or the maintenance of system performance) overlaid with a set of controls addressed to social problem solving, overlaid with another set performing monitoring or critical functions for the total system leading to active rather than reactive paradigm shifts? The innovative exploratory work in organization, reviewed above, does not come to grips with such questions, which surely are important to our understanding and practice of social learning.

Our discussion has suggested a more general way of defining the degree of entitivity of a social system than was developed earlier in the chapter. This principle would state that *the higher the degree of consonance between the social system goals and its controls, the higher the degree of its entitivity.* We are used to thinking of formal, hierarchical, prescriptive controls as yielding a high entity system because this organizational concept is based upon the machine system analogue and machine systems are high entity systems. The fact is, where such an organizational system is attempted in the face of great complexity (extensive boundary overlapping or extensive environmental change), the consonance between goals and controls turns out to be very weak. Consensus is forced or nonexistent. Such a system may turn out, in fact, to be a low entity system.

Conversely, one might be inclined to characterize as a low entity system an ecosystem with extreme boundary overlapping where controls are contextual and a great deal of subsystem independence prevails. Yet if such a system has an organizational form that matches the overlapping boundaries with overlapping controls and is supplied with a process that can generate a consensus, cohesion and system unity can emerge. The goals and controls of such a system might prove to be quite consonant and yield a high entity system. In the end the degree of entitivity of the social system and the consonance of system goals and controls rest, in turn, upon the level of awareness, participation, and consensus that characterizes the human components both as individuals and as functional subgroups. This is the fundamental source of social system cohesion and unity. The only alternative is consensus through coercion or propaganda—a spurious kind of cohesion that is dynamically unstable.

Communication as Dialogue. Critical to the organization of social learning is the understanding and practice of communication as dialogue. This is referred to by Matson and Montague (1967) as the "third communication revolution." They have this to say (pp. 1–3):

Just as we are accustomed to referring to the first and second industrial revolutions, so we may distinguish between the first and second communications revolutions. The earlier of these ... represented the triumph of scientific invention and mechanical engineering; it gave us, typically, the telephone, the radio, and the giant printing press. The second communications revolution is a triumph of scientific *theory* and *human* engineering; and it has given us, typically, cybernetics and mass motivation research.

... the cybernetic theory of communication ... is inherently *monological* and directive. For all its attention to the response mechanisms of feedback, its map of the communicative process is essentially that of a one way street. ... It is, in purest form, the model of the monologue—as opposed to that of the dialogue.

This point brings us to the newest (as well as, perhaps, the oldest) theoretical development in the field of communications—one sufficiently profound and pervasive to deserve the label "the third revolution." It is, simply put, the view of human communication as *dialogue*.

What do we mean by dialogue? Rappaport, writing in the Matson-Montague symposium (pp. 79–80), has this to say:

I believe the essential features of dialogue [to be]:

(1) The ability and willingness of each participant to state the position of the opponent to the opponent's satisfaction (exchange of roles).

(2) The ability and willingness of each participant to state the conditions under which the opponent's position is valid or has merit (recognition that any position has some region of validity).

(3) The ability and willingness of each opponent to assume that in many respects the opponent is like himself; that is to say, that a common ground exists where the opponents share common values, and each is aware of this common ground and, perhaps, of the circumstances which have led the opponent to the position he holds (empathy).

It should be apparent how this mode of communication is related to the organization of social learning. Monologic or directive communication will serve satisfactorily as a carrier for social system controls in a homeostatic system not excessively characterized by boundary overlap. Even here it can perform its social control mission if the human components of the system are either satisfied or compelled to conform to the control requirements. Where overlapping boundaries are common and the system faces problem-solving tasks in a changing environment, monological communication is not adequate to the tasks of integration and innovation. The attempt to rely upon prescriptive controls and monologic communication will, in time, throttle creativity and erode the consensus or sense of community that is the essential glue of the complex and changing social entity. The result is the destruction of both cohesion and adaptability.

The Likert study demonstrates this and reveals how these organizational failures are tied to (1) a system that inadequately accounts for the contribution of human resource components, and (2) organizational forms that inadequately promote and support dialogue. The Pelz-Andrews study reveals how important dialogic communication is in organizing the creative work of science.

The problem we face is this: we have been so absorbed with the monologic communication that controls the material transformations and transfers of society that we have paid scant attention to the design and creation of dialogic communication. In monologic communication a large part of the activity is addressed to the transfer of standardized signals over established channels, and the signals are generated by established criteria. In dialogic communication it is knowledge and behavioral designs that are both transformed and transferred. Two things are strikingly evident in dialogical communication. It is through the n-dimensional transfer and exchange of ideas that the transformation of ideas takes place—that creativity is spawned. The transformation-transfer dichotomy begins to lose its sharpness. Second, in the act of dialogic communication, channels do not remain fixed. They are often reshaped in the act of communicating.

In organizing the process of social learning we must obviously devote much more attention to the design and implementation of systems that make dialogue possible. It is through dialogue that developmental hypotheses are generated and the experiments in social system transformation are carried out. If social learning is to be made more efficient, less traumatic, and more responsive to the human needs, this is an area of major concern for social science and for modern society. This does not mean, of course, that all activities of the human components of social systems need to be guided in every act by a dialogical process. The dialogical process is the key to the agreement upon the context of the system—its goals and controls. It is only at the points where these contextual issues arise and where incompatible boundaries need to be reconciled that the dialogical process is essential. The day-to-day implementation can be governed by the more routine controls of monologic communication. But the human component can conform to these with ease if he understands and contributes to formulating the context out of which they arise.

Recognizing the significance of dialogue brings us back again to the interface between social system learning and individual learning and human development. We considered in the previous chapter the need for congruence between the requirements of social system organization and the requirements of individual human needs—particularly the need to support human growth motives. Beyond the necessity of maintaining animal existence, the human need focuses upon the meaning supplied by group unity and the exercise and development of creative potential. This discussion suggests that such a congruence is not only possible but is essential to fulfillment of the fundamental long-run objectives of both individuals and their social systems. We are beginning to understand that it is precisely through participation in dialogue that the peculiarly human needs are fulfilled.

We are now in a position to add further insight to the perspective of an earlier section (pp. 209 ff.), where we examined the way in which social learning modifies the social organization through its effect upon social system boundaries and networks. As social learning has taken place over the span of history it has successively modified the objectives and structure of social organization until it has brought us to the threshold we now face. Social learning has created the necessity of successfully organizing social learning so that it can serve human need. Understanding of the social learning process and the design of organization (goals and controls) that harnesses it to human service is the task to which social science must now address itself.

Social Systems and Social Science

Another new perspective relates to a point made in chapter IV. We can now see even more clearly how the special character of social science derives from the distinctive nature of the social system as an entity. As compared with physical and biological systems, the social system entity is less precisely bounded because of the greater prevalence of boundary overlapping and contextual controls. The problems of specification are multiplied. Not only is a given system more difficult to separate from overlapping systems, it is, by virtue of the overlapping and interdependence of components, more apt to represent a unique entity. If one adds to this the fact that social learning frequently redefines system boundary conditions, one can readily see that the problem of scientific generalization about social systems as entities is greatly compounded. This helps to explain why structural taxonomy, which plays a prominent role in the biological science, has played a much less prominent and useful role in the social sciences. It also helps explain the observation in chapter IV that law-like statements in the social sciences are more difficult to generate and less useful than in the biological sciences.

But there is a difference of kind as well as of degree. The goals and controls that define social system entity are a product of the consensus of its human components. The social scientist is a human being—a not fully determined system who acts as a component of the behavioral system to which he seeks to apply scientific analysis. Physical science tends to analyze exogenous systems from an external vantage point (notwithstanding the Heisenberg principle). Social science is self-analysis, and it cannot escape the constraints and obligations this places upon it.

Even physical science today is in a mood to see the need for self-examination and the need to anticipate the human consequences of its activities. (See Morison, 1969.) However, the effect of this is not to absorb social science into the physical science paradigm. The effect is to recognize that the practice of physical science must ultimately become subsumed within the social science paradigm.

Part IV

THE EMERGING
SOCIAL SCIENCE PARADIGM

VII   The Social Learning Metaphor

THE DEVELOPMENT OF THE THEME is now complete. It reflects two convictions: (1) Developmental problems are the paramount issues that confront social science and the social process. (2) The conventional and dominant metaphors of social science are, alone, not adequate to permit man to deal effectively with these problems.

We have examined the limitations of the conventional growth models and have concluded that the concept of learning is a more appropriate metaphor. We observed that a movement in this direction is clearly seen in social science, but that social science scholars have tended to settle upon development models that have to be characterized as programmed learning. There is a strong tendency to look at social change as if one were dealing with a machine system open to information or a deterministic sequence of development stages.

We concluded that only a creative learning model is adequate to cope with the reality of social and economic development. Pursuing this line, we examined the theory of modern synthetic evolution as a metaphor that might serve to enlarge the perspective of social science. While this model is highly suggestive and represents a decided advance over deterministic learning models, it is essentially a stochastic learning process. This theory concentrates upon the phylogenetic process by means of which the genetic pools of species are modified. It does not deal adequately with the way in which the learning organism which it spawned engages in the process of learning. It is especially inadequate to deal with the process of the only self-conscious, sym-

237

bolizing and symbol-communicating animal, man, engaged in learning in a social context.

The biological metaphor served us well, however, because it enabled us to focus attention specifically upon that aspect of social evolution that is unique. We became aware that the distinctive thing about the social process is that mankind, as individuals and as groups, is capable of *behavior directed to changing behavior*. Change is not purely stochastic, but includes a purposive element. We observed that this behavior-changing behavior embodies two other distinctive aspects of social behavior: behavior directed to evaluating behavior and behavior directed to organizing behavior.

We observed, further, that behavior-changing behavior conformed to a process that can best be characterized as evolutionary experimentation—the process of social learning. In this fashion we came to articulate a social learning metaphor distinct from biological evolution yet retaining elements of process similarity. Part III has been devoted to a detailed examination of this metaphor.

In Part IV it is suggested that this metaphor is beginning to form in social science. Its emergence is not complete or even yet assured, but it constitutes the emerging social science paradigm. The various disciplines of social science have gone through repeated shifts in the past as they have attempted to work, first, under the control of mechanistic metaphors, then the organic metaphor of biology, followed by the metaphor of biological evolution. It seems logical that the next step in the progression should be the shift to a social learning metaphor under the control of which social science can come fully into its own.

This is not to suggest that all of the disciplines in social science or their component subdisciplines move in a solid phalanx across each concept threshold. Much of social science research is still being carried out under the control of one or another of the more restricted metaphors. This is not inappropriate since there are many social problems for which these conceptual orientations are adequate to support useful work. Still, the more general of these models is coming increasingly into play as we find ourselves coping with developmental problems. There are many signs that what is characterized here as the process of social learning is beginning to emerge as a central paradigm for social science.

Part IV addresses this theme. In the present chapter attention is focused upon a summary of the nature of social learning followed by a discussion of some of the implications of this paradigm for social science. The closing chapter will attempt to show that many disciplines besides economics are finding different portals into the same conceptual domain and that the elements of the concept have been emerging here and there for some time.

The Nature of Social Learning: A Summary

The most striking thing about evolutionary history is that the operation of phylogenesis in its generalizing mode created improvements in organism adaptability until it generated learning organisms. Thus, the behavioral reprogramming of organism behavior that is phylogenesis gradually evolved a program (genotype) that provided the organism with the power to reprogram itself—to act as a true learning system at the organism level. In the human species the learning organism reached the point where learning becomes largely socialized because the dominant aspect of the individual organism's learning environment is the presence of and the sharing with other learning organisms.

In this way there emerged a process of social evolution distinct from biological evolution and a process of social learning distinct from a purely stochastic learning process. This process takes place at two levels—the level of the individual and the level of the group.

The major elements in an individual's learning and personality formation are associated with his social environment. He learns by sharing in a range of acquired information through communication and social system participation. Individual behaviors form group clusters to exploit the behavioral amplification that specialization and exchange can afford. Thus, individual learning and behavior are shaped by the nature of the group activities in which the individual participates and through which he is molded, and which in turn, are frequently molded by him.

Social learning also takes place at the level of the group, where the result is not the transformation of individual behavior but the transformation of group behavior. Either the collective membership of the

239

group or some leadership élite undertakes to monitor the group behavior with reference to group goals. Serious deviations constitute
problems which must be resolved through technological innovation
and social reorganization.

Social learning subsumes both social system learning and the socialized learning of the individual. Its operation at the two levels is
obviously interrelated. Since the orientation of this book is a social
science one, we have been predominately concerned with the way in
which the process works to transform social systems. The essence of
what we have learned about the operation of the process is as follows.

Social behavior is not identical with social learning. Two of the
unique aspects of the human social process continue to operate in the
absence of social learning. Behaviors directed to evaluating and organizing behavior are continuously employed in maintaining social
processes that are homeostatic in character. There they operate to
modify the mix and flow of established modes of behavior in response
to established behavioral criteria associated with the monitoring of
moderate environmental changes.

Since social systems frequently encounter situations in which such
predetermined forms of adaptation are inadequate, rarely can they
long escape the necessity for social learning. Then they must display
the third unique component of the social process—behavior directed
to changing behavior. In this learning process both behavior-evaluating behavior and behavior-organizing behavior continue to operate,
but they become transformed into aspects of behavior-changing behavior. They operate together to form a process of social learning.

This we identify as a process of evolutionary experimentation—a
process that is different in important respects from the experimental
process known to classical science. Social system evolution is the result
of a problem solving process that is an implicit if not explicit form of
hypothesis testing—but is problem solving and testing of a different
order.

Classical experimental physical science takes place at two levels:
analysis and system design. At the level of analysis it is concerned with
understanding the operation of deterministic physical systems from
the posture of an external observer. It commonly tests hypotheses
about the nature of the system by observing the effect upon compo-

nents of the system of changes in exogenous parameters, usually under highly controlled conditions. In this way it seeks to identify laws or universal relationships. At the level of system design, these relationships or laws are applied to the design of deterministic systems like machine systems, creating artifacts that amplify human behavior. The design of these systems is engineered from the posture of an external manipulator. To the extent that the valuation process is involved, it is concerned only with instrumental values or efficiency criteria at the design stage.

In social systems problem solving and hypothesis testing take on a different character. The basic point of departure is the fact that the social system experimenter is not exogenous to the system. He exists as an endogenous component of the system he is attempting to understand and transform. He is not dealing with the understanding and design of fully deterministic systems. He is immersed in the act of social system self-analysis and self-transformation. He is the agent of social learning—a purposive, self-actuating, but not fully deterministic process. He is not interested to the same degree in establishing universals because the social system which engages his activity is phenomenologically unique and both its structure and function are temporary in character. He is engaged, rather, in formulating and testing developmental hypotheses. The developmental hypothesis is a presupposition that, if the organization and behavior of the social system were to be modified in a certain way, the goals of the system would be more adequately realized. This developmental hypothesis is not tested repeatedly under nearly identical or controlled conditions. Rather, it is tested by the degree to which goal convergence is realized as a result of the experimental design. Problem solving—hypothesis formulation and testing—is an iterative, sequential series of adaptations of an adaptable, goal-seeking, self-activating system. It can be characterized as evolutionary experimentation.

This process calls into play the other two unique processes in its support. Because by its nature social learning involves social reorganization, behavior-organizing behavior is involved. We have seen that an organization is defined by a set of boundary conditions. These boundary conditions are a set of social system goals and the controls imposed upon component systems to assure that their behavior supports the total system goals. Developmental hypotheses arise out of the anomalies of social organization when the social system controls do not adequately serve the social system goals. They are tested through social system reorganization designed to bring the system goals and controls into consonance.

Evolutionary experimentation can be seen to exhibit both a normal **and extraordinary mode.** In normal problem solving the emphasis is upon the reorganization of the system controls. It involves the reorganization of subsystem boundaries under the control of the total system goals. It does not threaten the context of the system. As we have seen, this mode of problem solving is frequently inadequate. The developmental hypotheses and the associated subsystem reorganizations are not sufficient to bring about consonance of total system goals and controls. Thus, evolutionary experimentation occasionally manifests itself as a paradigm shift or a displacement in the operating context of the total system. It comes to deal, not with the reorganization of controls, but with the reformulation of total system goals as well.

This makes plain that behavior-evaluating behavior is also an essential servant of evolutionary experimentation. In the case of normal problem solving, attention is restricted to a consideration of the adequacy of instrumental goals and criteria. Subsystem goals may come under consideration because they are in conflict with the target goals of the total system. They are either placed under additional constraints by the reorganization of control subsystems or they are directly subjected to the reorganization of their own goals and controls. In this mode, however, even the revision of subsystem goals is undertaken under the control of total system goals. In the case of evolutionary experimentation that is manifest as a paradigm shift, it is the target goals of the total system that are subject to revision. This makes plain that the social scientist concerned with developmental problems cannot abstract from normative or value-laden issues. The process of evolutionary experimentation is essentially a normative process; one of its consequences is the evolution of social values.

Throughout social history the operation of this process of social learning has been hazardous and haphazard because its conduct has not been efficient and because it has been predominantly concerned with instrumental goals. In effect, the process of social learning has not understood itself sufficiently well to rationalize itself as an efficient process with a coherent purpose.

Because the process has not been understood and consciously applied, social change has frequently been dominated by an attempt to implement change by processes incompatible with the reality of social evolution. Acting on the basis of inadequate paradigms and metaphors, we have been inclined to practice a form of social engineering. It is presupposed that the change agents can act as though they are external to the process and have the knowledge and power to design a terminal state that will bring about consistent goals and controls in a deterministic fashion. Also acting on the basis of an inadequate paradigm of the social process, an attempt to freeze the social status quo—to create and maintain static social system boundaries—is frequently evident. Both of these impractical modes are clearly impossible, but the attempt to impose them upon social change has a tendency to exacerbate the traumatic and unpredictable consequences of the process.

The other part of the difficulty stems from the fact that social learning is being applied more efficiently in one of its modes than the other because it is only partly understood. The resulting disproportion has had serious social consequences. The conscious and efficient practice of physical science and the design of physical systems has been perfected to the point where normal social system problem solving is dominated by the reorganizations associated with new physical technology—the redesign of machine subsystems and their control. This has had the effect of accelerating the rate at which organizational crises occur—in particular, the rate of those that can only be resolved by paradigm shifts. This accelerated need for social reorganization is not matched by an appropriate understanding of social system learning. We do not yet see the implications of normal problem solving for the control of human subsystems (involving the consistency of individual and total system goals), nor do we see clearly how to engage purposively in the process of controlling paradigm shifts. This has led to an accelerating rate of painful, reactive, major social reorganizations. The revision of the target goals of social systems is as yet largely uncoordinated, fragmentary, and opportunistic.

One concludes that amelioration of many of the world's worst social ills, if not the long-run survival of the social process itself, must hinge upon our ability to make the practice of social learning more orderly and rational.

First, we need to devote concentrated attention in social science to understanding the process. Second, at every stage and level of our understanding we need to apply what we know to conscious, orderly practice and control of the process. This implies that developmental hypotheses should be more objectively and consciously formulated by the group. The evolutionary experiment should be frankly conceived as an experiment and deliberately provided with information feedback that monitors goal convergence and sets the stage for the next round of experimentation. The seductive appeal of utopian social engineering must be put aside. Third, we need to innovate organizational forms and procedures that efficiently integrate the goals and controls of social learning itself. Fourth, we need to acknowledge that this may require an over-arching social goal or value that serves as a final test for evolutionary experiments—that guides the formulation of developmental hypotheses and passes judgment upon paradigm shifts.

We have suggested that this over-arching goal is the development of the growth motives of the human individual. If this suggests that, as a consequence, social evolution becomes teleological in a way not matched by cosmological or biological evolution, it should be pointed out that this is not a terminal state or a transcendent teleology of the kind associated with the orthogenetic fallacy; nor is it the conventional teleology of social action formed by the pursuit of instrumental goals and social maintenance. It is a process teleology. It suggests that human beings can establish the process of human development as the goal of the process of social evolution. Both the process and the goal are understood to be open to further transformation as we advance in the practice and understanding of them.

By asserting the normative priority of human development as our social goal, we add to the internal consistency of the social learning paradigm as well as to the order of the process. It provides the interface between the social learning of the individual and social system learning. It is only through establishing such a priority that group behavior and individual behavior come to be mutually reinforcing in a nonstatic world.

If this makes the social process and social science anthropocentric in

244

character, no apology is necessary. To deny that this is appropriate would be to deny a fundamental aspect of human nature and the evolutionary process that formed it. The psychic orientation of man and his motivation to action is by nature anthropocentric just as that of the rat is rat-ocentric. Man happens to have the potential through social learning to create a controlled process that can support him in the realization and exercise of his highest human potential—perhaps even to enlarge that potential and the meaning and joy that accompanies its exercise.

Implications of the Emerging Paradigm for Social Science

The essential features of the emerging social science paradigm have been detailed. Social learning is shown to be a process of evolutionary experimentation that constitutes individual and group self-analysis and self-transformation. It incorporates activities engaged in evaluation and social reorganization. If this metaphor is adopted by social science to provide the context for its work, a reorientation in a number of its views and practices is implied. This section examines several of these implications, some of which have been mentioned earlier in a different context, and others are considered here for the first time.

THE ROLE OF ORGANIZATION

The attachment to less general paradigms in social science than the one we are discussing has led to a rather restricted view of the study of social organization. Attention has been devoted primarily to ways in which social organization serves social maintenance. The study of organization has meant, essentially, study of the organization of homeostatic systems. Organization is related to social change in the sense that it provides for management of preestablished adaptations in the face of marginal parameter changes, with the main emphasis upon the organization of a system throughput. Social engineering is seen as the design of optimal static state networks. Minimizing the costs of the transfers and transformations under the control of the system is stressed

245

rather than the nature of system goals and controls. The latter are essentially taken as given. A vast bibliography could be assembled of social science studies and organizational plans that conform to this description.

When one begins to operate under the control of a social learning paradigm the role of organization and the nature of its study are extended in important ways. Then social reorganization becomes revealed as both the target and the instrument of social change. This orientation requires social science to consider much more seriously and more fundamentally than has been its habit the sense in which a social system is an entity. It needs to understand that a social system is not just a spatial or functional agglomeration of social activities or just a functional network of interrelated activities. This has been the conceptual focus of much of the social science study of organization. It needs to focus upon the boundary conditions that define an organization and to see the social system more clearly in terms of the goals and controls that shape its behavior. Two reasons have become apparent why this reorientation is important.

First, social organization is the target of social change. It is the anomalies of social organization in a changing world that constitute the motive and object of social learning. The anomalies are manifest in the discrepancies between social system goals and the operation of social system controls. The formulation of evolutionary hypotheses is the visualization and choice of a procedure that it is presumed would improve the consonance between goals and controls. The evolutionary experiment is an experiment in social reorganization or boundary redefinition. Both the understanding and the practice of social learning require that organization be seen in this light.

Second, social organization can become the instrument of social change. Social science needs to become more concerned with the role of organization in providing goals and controls for the process of social learning itself, and with the forms that it might take in performing this role. It needs, specifically, to inquire into the differences in form that might be appropriate for the management of normal problem solving and paradigm shifts.

Associated with these two points is a third, which has been emphasized earlier. Conventional treatments of organization have tended to concentrate on the structure of networks and their efficient design. This study of networks deals with the physical and symbolic transfers that link component activities in a social context. Since the controlling metaphor has been the equilibrating or homeostatic one, attention has focused on the development of theories of transportation and communication as one-way transfers and on the design of transfer systems as efficient mechanistic social subsystems.

When we are faced with the reality of the social learning process and work under the control of its metaphor, it becomes evident that since social systems are frequently temporary systems, less emphasis needs to be given to the nature and design of optimum transfer networks and more to the design of adaptable networks—systems that can more easily be adapted to the requirements of new social goals and controls. If this could be done, a contribution would be made to reducing the capital costs of change. More important, we become aware that the communication that serves the reconciliation of social goals and those of the human components of social systems, and that serves the process of social learning itself, is not a monologic or one-way transfer process. It is a two-way, dialogic process. Social science needs to concern itself with the theory of dialogic communication and the design of dialogic social organizations and processes.

THE ROLE OF HISTORY

If the central concern of the emerging paradigm is with process, then social science must accord to history a more significant role than it has heretofore played. That role may take on a character that appears novel by more traditional standards. If we view social evolution as the product of a process of evolutionary experimentation and valuation, then history must be a record of that process, and it must condition at each stage the developmental options or developmental hypotheses that are available to the social experiment.

If history is viewed in its supporting role in the study and practice of social learning, it is not particularly helpful to approach it in a more traditional context—as a repository for anecdote or chronology, as so much of the historical record is conceived. Nor should it be viewed as a manifestation of a theory of history, as so much else of

247

the developed historical record is conceived.[1] The social learning metaphor does not see the role of history as the interpretation of the historical record in the light of a theory of history. It sees the hypotheses of evolutionary experimentation as subject to testing and reinterpretation in the light of the record of history. This is quite a different orientation!

History, interpreted in this way, plays its role at two different conceptual levels and in two distinct temporal orientations.

At one level we visualize the operation of the process of evolutionary experimentation in social systems as they seek the progressive solution of perceived social anomalies. In this context it is the role of history to provide the information essential to the process. The relevant history and the information it provides is divided into the past and the future. From the point of view of the past, the study of history joins the study of organization in revealing the nature of social anomaly.[2] Such historical materials are important in defining the social problems to the solution of which evolutionary experimentation is directed. They yield the premises for developmental hypotheses.[3]

From the point of view of the future, history provides the information that reveals the performance of the system under the influence of the social experiment and generates the information base that provides the test of the developmental hypotheses in terms of the presence of goal convergence.

At another conceptual level, we visualize the emerging social science

---

[1] Both implicit and explicit theories of history are legion. Some amount to no more than the naive interpretation of history as the record of the influence of great men. Others see history as a manifestion of a social life cycle whereby social systems go through a cycle of birth, maturity, decline, and death similar to the ontogenetic cycle of the biological organism. No lesser figures than Plato, Spengler, and Toynbee have devoted substantial parts of their careers to such historical interpretations.

[2] In doing so, it plays a double role. It provides information concerning the current state of social system boundaries and the additional perspective of their "coming-to-be."

[3] The historical statements and materials developed in this context may make use of a background of general historical information, but they pertain primarily to the historical experience of the social system subject to behavioral analysis and transformation. That is the focus. As such, the historical statements developed correspond to what Goudge (1961, pp. 70–79) characterizes as "narrative explanations."

paradigm itself. Here we need to recognize that all evolutionary theory, including the concept of social learning formulated here, is a set of historical statements.[4] The social science paradigm that is emerging is supported by a large, steadily increasing body of historical evidence. This suggests that the testing, refinement, and enlargement of the paradigm itself is in large measure a function of historical scholarship. As before, the role of history has an orientation to both the past and the future.

From the point of view of the past, we need to bring historical materials to play in testing and expanding our still primitive understanding of the process of social learning. The emerging paradigm is incomplete and perhaps partially in error. We need to study historical materials intensively in order to identify the way in which evolutionary experimentation has worked to bring about social reorganization down through the ages. Especially we need a much better record of the evolution of social goals and controls, and also of the circumstances and conditions that have attended successful paradigm shifts in the form of social reorganization and scientific evolutions. We understand the normal process of evolutionary experimentation under the control of an established context much more adequately than we understand the extraordinary processes associated with paradigm shifts, yet so much of our present concern with development problems in social policy involves the processes associated with paradigm shifts rather than the normal process of social change.

It is not suggested here that the process of social learning be taken as a "theory of history" to be applied to the interpretation of historical materials, rather, that the provisional view of that process be tested in the light of relevant historical materials so that irrelevant portions may be falsified and relevant additions to the concept revealed.

The object of such historical research is not to yield a *completed* conceptual model of the process of evolutionary experimentation. We have enough evidence from history to know that the process itself is evolving. What we seek is a better understanding of the evolution of

---

[4] They correspond to what Goudge (1961, pp. 65–70) refers to as "integrating explanations."

the social process and as complete an understanding of its present state as possible.

From the point of view of the future, because of the hazardous and inefficient nature of the process of social learning reflected in past history and current experience, the emerging experiments in the organization of social learning will have to be performance tested by an emergent future history.

In closing these comments on the role of history, a methodological note may be in order. The historical statements that reveal the nature of the process of social learning and form the test of its successful application are different from the scientific statements yielded by classical experimental science. They do not form predictive laws that define the universal characteristics of a class of phenomena.[5] The laws of classical science support positive prediction within their domain of validity. These historical statements do not. This does not make them or the process that generates them less scientific. The test of a scientific statement is not its power of positive prediction but its susceptibility to falsification. These statements are subject to falsification and revision by historical evidence.

More important, such statements have a predictive power of a different kind. They do not support a positive prediction but they imply a negative prediction—otherwise they would not be subject to falsification. In short, evolutionary theory will not allow us to say, "If we do behave thus, the following will happen." But it may permit us to say, "If we do not behave thus, the following cannot happen"—with no implication that such necessary conditions are sufficient. This is just another way of saying that the evolutionary process is not deterministic and, therefore, its study will not yield to the methods of classical experimental science. What we can hope for is that, through the study of the process of evolutionary experimentation and through the conscious objective application of this more generalized scientific method, we can refine our negative predictions. We may bring the understanding and exercise of the process to the point that we will be able to rule out a larger number of developmental hypotheses before be-

[5] This should not be interpreted as saying that no "law-like" statements can be derived from historical materials.

havioral testing because they are seen not to satisfy the necessary conditions of the process. This is the task of social science.

THE ROLE OF SOCIAL ACTION IN SOCIAL SCIENCE

The view of the social process as one of evolutionary experimentation in the pursuit of goals modifies the social scientist's relationship to social action in two ways.

Let us first consider his role as a student of social action. As long as he is studying the social action characteristic of homeostatic systems[6] the supposition is that he is dealing with stable social structures and goals. The focus is upon those activities joined in the management of the equilibrium system and the refinement of its efficiency, or upon the displacements of homeostasis that follow directly from changing dimensional parameters (like population size, income, etc.). Once the social scientist views social action as a social learning process, he must become concerned with the study of social goals as they relate to the process and study of organizational forms, already discussed, that are appropriate to adaptability as well as adaptation. He must also take a different approach to the information resources of social science.

There is a second way in which the social learning metaphor changes the relationship of the social scientist to social action. He can no longer abstract himself from social action when it becomes the evolutionary experimentation by means of which social systems pursue and modify their goals. The social scientist can no longer emulate the traditional practice of physical scientists in this matter. He can no longer restrict his role to that of supplying information and techniques to social system managers. He has to become more than a consultant in the "social efficiency game."

He must now play a role in the study and practice of evolutionary social experimentation. It becomes an important part of his role to monitor the discrepancies between social system goals and controls—i.e., the identification of social problems—and to identify and present

---

[6] Or what cybernetics would call an "ultra-stable" system—i.e., one in which a shifting homeostasis takes place as some of the dimensional parameters of the system become modified.

the behavioral options one might choose in formulating the developmental hypothesis. This requires both an objective study of the behavioral range of social learning systems and the evaluation of the revealed options in terms of their consistency with the developmental history of the system. His role also requires him to guide the reality testing of the social experiment, to propagate an awareness of the techniques and constraints involved in the practice of social learning, and to train social actors in the appropriate skills.

These are not roles he can play on the sidelines of the political process that carries out evolutionary experimentation. The contribution that he can make to the rationalization of the process of social learning cannot be fully carried out without becoming an actor in the process itself. The emerging paradigm seems to require that we develop a new breed of politically oriented scientists and scientifically oriented politicians—that social analysis and social action be merged.

THE ROLE OF INFORMATION

The above discussion makes plain that the social learning model implies a different role for information in social science. Let us examine this point briefly.

The logical point of departure is the realization that all behavioral systems, biological or social, are information or knowledge systems. They receive a continuous flow of information about the states of the environment. These are filtered through the boundary conditions of the system—a symbol system or design that interprets the meaning of the environmental signals in terms of the system's goals. These symbol systems or behavioral designs are, themselves, a form of information or knowledge. If the meaning attached to environmental information suggests a behavioral response, the system boundary generates a set of control signals that inform the component systems how to behave. It monitors the appropriateness of the response by gathering new information about the state of the environment, and so it goes.

This perspective of the behavioral system as an information system enables us to compare the role of information under the control of the mechanistic or deterministic metaphor with its role under the social learning metaphor.

First, take the concept of the homeostatic or equilibrating system. Under the control of such a metaphor the behavioral design implicit in the system boundary (i.e., the goals and controls) is taken to be given. It is treated as stable and static. In these circumstances those participating in system management are only interested in those forms of information that reveal the state of exogenous and endogenous environments that require behavioral adaptation under the control of the paradigm and inform the controls essential to the adaptation. The business manager, for example, is primarily concerned with information that reveals the conditions of the markets in which he buys and sells and the performance levels of operating divisions and departments.

In this case the social scientist, acting as a technical consultant to management, seeks essentially the same information. At most, he is concerned with how this information can be interpreted under the control of the mechanistic metaphor to reveal some technical improvements in control efficiency. But the social scientist may play another role. He may seek to understand the larger ecosystem of a set of interrelated social systems of this type. For example, this is certainly the character of the economist's interest in the iterative coadaptations of the market mechanism. Since this ecosystem does not have a set of explicit boundaries and controls, he is concerned with identifying and making explicit the implicit boundaries, i.e., with understanding the system.

At this higher order system level the social scientist feels little interest in the internal information systems or the control processes of what now constitute the subsystems. He is content to treat these as entities. He develops and works with crude taxonomies of such entities (e.g., the national industrial classification code). The information he seeks is dominated by the transfer relationships (input-output) these entities form with other entities and the dimensions of the resource and capital stocks that generate these flows. From a different perspective, he is employing the same kind of information that the system manager seeks about his external environments. It is general market, demographic, financial, or other such information cross-classified with the conventional social entity taxonomy. Standard sources of information based upon standard taxonomies tend to become dominant. The information tends to flow as a by-product of the operation of the control monitoring functions of the system. Organized reporting systems emerge that marshall these standard forms and sources of information.

Because the homeostatic paradigm has been the dominant one in social science, this is a fairly accurate representation of what has, in fact, been the dominant role of information in social science in the recent past.

Now consider the different role of information under the control of the social learning metaphor. Social learning is concerned with the revision of social system boundaries. It is explicitly concerned with the adequacy and reformulation of social system goals and controls. That which is taken to be given under deterministic metaphors becomes the focal point of analysis and practice. It follows that those forms and sources of information that have been traditionally exploited by social science do not adequately serve the study and practice of social learning.

What is needed is information about social system goals and controls that will reveal the degree to which the system's response to environmental signals is goal satisfying. The identification and definition of social problems require the ability to judge the consistency of social system goals and controls. Once the problem areas are identified and the need for boundary revisions established, information is needed about the behavioral options that are candidates for formulating a developmental hypothesis. This requires a means for becoming informed about the goals and controls and technologies employed by other systems that might be borrowed as well as inventive imagination. Once a developmental hypothesis is formed, one needs political information concerning the system's human constituents necessary to the formulation of a consensus. One needs a technology of organization for change. Once the social experiment is performed, one needs information that measures goal convergence and, hence, is necessary for reality testing.

Here we are talking about information about social system goals, the goals and attitudes of human components, machine technologies, organizational or control technologies, and similar matters. None of these are forms of information provided by the classical information systems of social system management or social science. The conventional internal accounting and reporting systems of management and the traditional general purpose, general parameter information systems conventionally employed by social science do not generate information of these kinds.

This implies that the entrepreneurial leaders of social learning systems, and the social scientists who want to both study and participate in the process, will need to develop new procedures for generating and handling these neglected forms of information.

Consider, first, their generation. The static taxonomies and conventional reporting techniques presently so characteristic will not be adequate to generate these kinds of information. We are already finding out that, in the face of substantial and rapid social change, traditional taxonomies (like the standard industrial classification code) yield information of considerable ambiguity even for conventional analysis and management. The taxonomies tend to lose touch with the realities of social system structure. Clearly, general purpose taxonomies will lose their importance. The emphasis on special purpose information will increase. We will be faced with problems of monitoring the behavior of complex, living social systems and at the same time making sure that the information process itself does not destroy or distort the aspect of system behavior we seek to observe. We may, like the physician, have to invent subtle tracer techniques. The need for information about goals and controls will generate information that could be used to violate or inhibit fundamental humanistic goals in a coercive social control system. This raises the need to devise special safeguards and protections. Some of the information will have to be generated out of dialogue to meet a given experimental situation.

Consider, next, the processing and organizing of information for use. Three things seem clear:

(1) The old techniques for storing and retrieving information for general use are inadequate for this new requirement. For numerical data the standard technique has been to organize general purpose statistical materials into intermediate aggregates based upon some standard classification. These were supposed to serve a wide variety of uses because the taxonomy is supposedly related to stable artifacts and social entities relevant to many forms of useful inquiry. Retrieval of

255

such information from repositories and its organization for use are simply achieved by manipulating these aggregate or summary data. Once we deal with behavioral systems that are engaged in evolutionary experimentation, each situation often requires a different functional typology and different ways of *relating* information in analysis. General information storage and retrieval systems will have to be designed with a great deal more adaptability and taxonomic flexibility. (See Dunn, 1965.)

(2) Techniques for the efficient generation, storage, and retrieval of textual and other symbolic materials (besides numerical data) will have to be worked out. The importance of goals, values, and other qualitative forms of evidence will require ingenuity in the generation and use of information.[7]

(3) A third implication is that special purpose information systems will have to be given a great deal more emphasis. Much of the information required by the process of evolutionary experimentation is peculiar to the operation of the system engaged in the process. But whether this information can be drawn from common sources or is self-generated through the system's own behavior, *its organization to support the process of social learning through evolutionary experimentation requires that the information be organized into what, these days, is referred to as a "realtime" information system directly related to the experimental process.*[8]

The social learning metaphor requires not only that the social scientist make a substantial reorientation if he wishes to work under the control of its context, but also that he substantially modify his attitudes

---

[7] Implied in this whole discussion also is the need for developing new analytical methodologies as well as new methods for generating and managing information. For example, the new mathematics of "fuzzy sets" seems to be a move to innovate in this direction. It is designed to deal with qualitative and even verbal parameters that can be introduced into the deductive logic that serves analysis.

[8] Realtime information systems are those that are organized for the purpose of serving a control process directly. They tie environmental sensing, interpretation and behavioral response in a continuous feedback loop. Homeostatic systems are typically well organized realtime systems. The problem we face is one of converting social learning into a controlled realtime process. This requires the development of a realtime information system that can serve this purpose. For a further discussion of realtime systems as they relate to this process see Sackman (1967).

about and interest in information. This will place quite a different burden of responsibility upon the social scientist and calls for quite a different allocation of his professional effort. The social scientist has never developed the tradition of responsibility for his research information that has long characterized the experimental physical sciences. For the most part, he has been content to utilize conventional general purpose information sources. The attitude that data generation is "dirty work" preferably left to someone else has been common. Among other reasons, this probably stems from the fact that social science has not generally been thought an experimental science.

But under the control of the emerging paradigm, social science begins to emerge as a genuine experimental science in a more general context. If it is to succeed in carrying out the responsibilities of this role, it must accept a new responsibility for the generation and application of relevant information. During the early phases of the practice of this new science, substantial resources and intellectual effort will have to be devoted to creating new information sources, methods and systems.

THE ROLE OF INTERDISCIPLINARY RESEARCH

The fifth implication of the emerging social science paradigm is the fact that social learning under its control requires the cooperation of all of the science disciplines. One might go further and suggest that it provides a framework within which the sciences may, in time, become truly unified.

Such a merging of the disciplines would be significant on three levels. First, it implies a joint role for all the social sciences. They all have a necessary contribution to make in specifying the nature and pursuing the exercise of social learning. No single social science discipline is adequate to take on the task. They all can find a common purpose and orientation within this paradigm. The interdisciplinary work required to understand the interface between social system learning and the socialized learning of the individual is an example.

What may take place here is similar to the rapprochement that is already underway between the physical sciences and the life sciences.

257

Simpson (1963, pp. 87–88) illuminates the process that we suggest must take place in the social sciences.

In our own days, Einstein and others have sought unification of scientific concepts in the form of principles of increasing generality. The goal is a connected body of theory that might ultimately be *completely* general in the sense of applying to *all* material phenomena.

The goal is certainly a worthy one, and the search for it has been fruitful. Nevertheless, the tendency to think of it as the goal of science or *the* basis for unification of the sciences has been unfortunate. It is essentially a search for a least common denominator in science. It necessarily and purposely omits much of the greatest part of science, hence can only falsify the nature of science and can hardly be the best basis for unifying the sciences. I suggest that both the characterization of science as a whole and the unification of the various sciences can be most meaningfully sought in quite the opposite direction, not through the principles that apply to all phenomena but through the phenomena to which all principles apply. ... they are the phenomena of life.

In short, the accumulated knowledge of all the social science disciplines can find application within the framework of the social science paradigm—the phenomena of social learning. Furthermore, acceptance of the obligation to work within such a framework will reorient the research content of every discipline and have the effect of blurring the lines of demarcation that now mark them. Cross-disciplines of the kind already appearing in the physical and life sciences (e.g., biochemistry, biophysics, etc.) are beginning to appear in the social sciences and will undoubtedly become more important.

At a second level the social science paradigm suggests a rapprochement between the social sciences and the physical sciences. If the learning systems represented by ontogenesis and phylogenesis constitute a paradigm under the control of which the physical and life sciences can converge, then the paradigm that is social learning must offer the framework within which the social sciences, life sciences and physical sciences can all find a relevant place. One cannot conceive that the exercise of the process of social learning can be understood or can be brought under purposive control without the active participation of all sciences. The task of making the process less hazardous requires a greater integration of these efforts.

The necessity for and nature of this rapprochement have been anticipated elsewhere. In an eloquent statement Burtt (1965) has this to say:

Every act of a scientist in his role as a scientist has varied and complex effects. Traditionally, he has restricted his attention to the effects of the physical world, for it is there that he looks for the outcome of an experiment or a set of observations that provides an answer to the question he has asked. Other effects, which do not belong to that world, he has excluded from the realm of science and from the scope of his responsibility. (p. 188)

[But] . . . in today's world the physical and social consequences of scientific activity cannot be separated. They form an organic whole, and all of them alike fall within the realm of our capacity for predictive understanding. There is not, and cannot be, any such thing as 'pure science,' if we mean by that phrase a form of scientific activity that has no foreseeable consequences. (p. 186)

He points out that the consequences of this view are far-reaching.

. . . the consequences would be pretty radical. The consequence for the detailed structure of science would be that it would then include the many kinds of predictive thinking it has thus far left out—at least all that show a conscientious concern for relevant evidence. The general consequence, however, would be still more radical, ... When we ask what would give science its organic unity in that enlarged structure, the conclusion seems unavoidable that all science, physical and social, would then be unified under the principles gradually taking form in the latter. The reason is twofold: (1) the attempt to forecast the effect of scientific activities on society is clearly a part of social science; and (2) the periodic reconstruction of scientific presuppositions under the influence of newly emerging interests also belongs to social science, so far as the process can be predicted by wise insight. The physical sciences, without losing their distinctive character, would thus become a branch of the science of man. It is the task of the latter to understand all that men do in their interaction with the environing universe, including their adoption from time to time of whatever presuppositions seem promising for explaining the physical world. The ultimate key to all predictive understanding, however strange it may sound, is human understanding.

In short, the whole fabric of our predictive knowledge may fulfill its promise only when it is unified under a dominant interest that at present might easily appear irrelevant. (pp. 175, 190)

259

Briefly, the emerging social science paradigm must involve both the physical and social scientist in a wider view of their roles. The domain they face is more difficult than previously addressed and their ingenuity and commitment to the habit of truth will be taxed as never before. But if man makes any pretext at either understanding or fostering and controlling social development, this task must be engaged. All the most vital stakes of society are involved.

Moreover, there is a very poignant reason for accepting that obligation; whether the scientist himself will have a future is at stake. ... Simply and starkly put: the adoption of one conception of science will tend to create the conditions under which it can continue to exist and grow, while the adoption of another (or rather the preservation of an outmoded one) will tend to destroy those conditions. (pp. 186–87)

At still a third level, the emerging social science paradigm suggests a rapprochement between science and philosophy. If we accept the fact that the process of evolutionary experimentation is, at the same time, a process of progressive valuation, and if we accept that the process of revaluation that is inherent in major paradigm shifts (both in scientific concept and social organization) needs to be tested by the target values of a higher order paradigm, that higher order paradigm must ultimately include the contribution of speculative and normative philosophy. But it is a philosophy that will have to be conducted differently from the past, as Adler (1965) points out. (See chapter VIII, pp. 270 ff.) It will need to be conducted as a conscious learning exercise consistent with the process of social learning—employing scientific information as the raw material for its speculation, using scientific information to test its speculations, and assiduously searching for the social anomalies that reveal inadequacies in philosophical paradigms and suggest their progressive revision. It is an exercise that will have to be conducted by philosopher-scientists and scientific philosophers. Allport (1955, p. 17) anticipates this when he says: "The goal of psychology is to reduce discord among our philosophies of man ..."

For a long time the history of the development of science has followed a pattern of subdivision and divergence that has been such a common manifestation of the evolutionary process. One can trace a kind of "phylogenetic tree" of the sciences branching out from its roots in philosophy. We now appear to be commencing a phase of diffusion and convergence. It is quite possible that the process will reconverge upon philosophy, its original home.[9]

Epilogue

Two closing observations are worth making. First, the social learning paradigm is, itself, offered as a hypothesis subject to testing. Second, the social learning paradigm may forecast a fourth threshold in human-social evolution.

THE PARADIGM AS HYPOTHESIS

We entertain no illusion that the representation of the social learning metaphor offered here is complete or final in any sense. It is without a doubt subject to further articulation and refinement. It may well contain error as stated. It may come to be displaced by another concept model deemed superior. In any case, this paradigm is in an early emergent phase. If it has merit, the bulk of the credit is due to the host of antecedent scholars upon whom we have drawn so heavily. Our attempt to draw these materials into a more comprehensive statement may occasionally be characterized by unjustifiable license or illogical inference, but we believe that its main characteristics are sound.

In developing the concept we have found it necessary to take a brief excursion into philosophical speculation. We have also made a commitment of faith to an, as yet, inadequately tested paradigm. This may be disturbing to some who still cling to the belief that the purity of science is compromised in this way. For this we offer no apology. It seems obvious to us that no general metaphor that controls the conduct of science can be characterized without engaging fundamental philosophic presuppositions and without a conditional act of faith in that metaphor.

In effect, what we are offering here is a paradigm hypothesis. It is subject to evaluation and testing in the same way all paradigms are. It will become established when a substantial consensus develops in

---

[9] An observation also made by Boulding (1961).

the social science community that this paradigm is superior to any now existing. It will achieve this consensus if it is logically more defensible and/or proves to be superior as a guiding framework in the pursuit of social science (which would now include social system experimentation). It will stand as long as there is no superior alternative. It will be displaced if another comes along that is logically superior and proves more useful as a control framework for the conduct of social science. This is the only kind of test to which a conceptual paradigm can be submitted. It is to such a test that this representation is offered.

THE FOURTH HUMAN THRESHOLD?

We close now with a question not as yet answerable. *Is it possible that, through the conscious, objective exercise of the process of evolutionary experimentation, mankind may approach and achieve a fourth human threshold?*

The reader will recall that in chapter III we identified three major thresholds in human-social evolution. The first threshold occurred when phylogenesis operating in the mammalian stem raised the generalizing capacity of the animal nervous system to the point where it could support symbolization and, hence, speech and other forms of communication. This led to the emergence of socialized human learning. In the early phases of social evolution human learning was dominated by the sharing of information and behaviors having to do with the individual's relationship to the nonhuman environment. The adaptations sometimes spread rapidly because of the demonstration effect and human communication, but the adaptive behavior was not, itself, highly integrated or socialized in nature.

A second threshold emerged as the social organization of behavior addressed to social maintenance progressed. The organization of the physical transfers and transformations that mark the everyday task of maintaining human populations gave rise to social systems with communication tendrils (along with their symbols of language and numbers) and transportation linkages. Men as individuals became progressively immersed in a social environment made up of the behaviors and responses of other individuals *and social groups*. The behavioral amplification associated with social organization permitted improvements in efficiency which, in turn, permitted improvements in the levels of biological and physical welfare and/or increases in the sizes of supportable populations. This was achieved by a process of unconscious and undirected evolutionary experimentation. It was slow and inefficient, requiring millenia (and then centuries in a slowly accelerating tempo) to work its changes.[10]

The process of social learning that yielded the emergent social changes was not understood or deliberately employed in the process of human problem solving. Innovative ideas were individually inspired and generally lost in the inertia of a predominantly static state social process and social mentality.

The third threshold was crossed quite recently. This was achieved when man developed a conscious understanding and application of a significant component of the process of evolutionary experimentation—the classical scientific method addressed to the understanding and design of deterministic systems.[11] Man came to understand systematic empirical testing as a powerful method for advancing useful knowledge about the material universe. Empirical experimental procedure emerged by fits and starts during the Middle Ages but became a widely understood and practiced procedure only during the last three hundred years. The principle found application through experimental machine system design and fed the dramatic technological and industrial revolution.

This, in turn, forced major changes in social organization and behavior necessary to exploit the potential for behavioral amplification this new technology offered. But the adaptation of individual and social behavior that resulted was still carried out by a largely ex post and unorganized process of piecemeal problem solving almost as old as human history. This has created serious anomalies in the social process. The old process of social reorganization and social valuation cannot

---

[10] Slow and inefficient, that is, from man's twentieth century perspective. Sociogenesis has been remarkably efficient and rapid when compared with phylogenesis.

[11] There were earlier thresholds such as the development of mathematics, logic, and axiomatic reasoning. Their emergence was an essential precondition for the development of the scientific method and their discovery contributed to the gradual acceleration of social learning. But the critical threshold appears to have been the scientific revolution.

keep pace with modern physical science and the major changes (largely unanticipated) its application has forced upon mankind. It is further complicated by the fact that the social reaction to these anomalies tends to take two forms. On the one hand, greatly impressed with the achievements of technological science and misunderstanding the nature of its method, many are led to believe that a form of human engineering and large-scale holistic social redesign is possible. On the other hand, many are tempted by a withdrawal reaction that sees a major threat in all change. The behavior of the utopian and the stagnationist has vastly exacerbated the problems associated with this phenomena.

It is this critical state of affairs that is breeding the widespread movement that considers social change as a manifestation of a more general learning process—as has been reported in this volume. A science paradigm is emerging that sees classical experimental method and classical science as inapplicable to the more general task of social problem solving. It is limited because it does not deal adequately with the problem of valuation and because it is inapplicable to realtime experimentation in the context of an operating system. The emerging social science paradigm is trying to make manifest a more general and inclusive process of evolutionary experimentation that is applicable to all social problem solving.

This effort is in an embryonic stage and it is too early to be assured of its success. If it could succeed, mankind might at last cross a fourth human threshold where he acquires the power to shape *behaviorally patterned environments* that serve human development in both the physical and social domains. He would no longer need to submit to *environmentally patterned behavior* inimical to his human potential. This is a heady prospect. As yet we can catch no more than a glimpse of the world that lies beyond that threshold. But one thing is certain: those of us who have a commitment both to science and human values have a great deal of work to do.

VIII   Optional Portals to the Same Domain

The summary discussion of chapter VII marks the end of the development of the theme of this book. The reader can stop at this point without feeling that the argument is incomplete. The concept itself, of course, is not complete by any standard: how much we have yet to learn has been repeatedly emphasized.

It seems desirable, however, to extend the theme of Part IV by bringing more evidence to bear upon the notion that the social learning metaphor constitutes an emerging social science paradigm. One can observe the elements of the metaphor visible in the work of a number of scientists in a variety of disciplines. It is useful to realize that there are many portals into this same domain of concept and that most of them have already begun to open. It may enlarge our understanding of the concept to see it from several different perspectives.

For another reason it is especially appropriate that we do this. In the preparation of this volume a deliberate choice was made not to burden the development of the concept with extensive notes and cross-references to the literature. It was an expedient adopted to aid the exposition of a topic inherently difficult to deal with. No effort was made to satisfy the canons of traditional scholarship. Accordingly, the reader may be left without an adequate sense of the great debt the author owes to many creative writers, or of the great ferment to be seen in science these days as it reaches out for new concepts to guide its work. The present chapter may add this perspective.

Relying chiefly upon recently published works, an attempt will be made to illustrate how elements of the social learning metaphor are making their appearance in such diverse fields as science history, philosophy, psychology, general systems theory, decision and organization theory, economics, sociology, political science, and the study of technology. Because the Marxian theory of social change has had such a great historical impact, it also seems useful to see how it compares with the social learning metaphor. What follows takes on the form of a collage of concepts designed to leave some impression of the broader intellectual context within which this work rests.

Science History

The thought-provoking work of Kuhn (1964) has already aided us by supplying the distinction between normal problem solving and paradigm shifts. That work also illustrates another portal into the domain of the social learning metaphor.

Kuhn writes as a science historian and the study that details his view of the way science progresses was a product of his dissatisfaction with the prevailing notions in the scientific community about the nature of science and the general practice in chronicling the history of science. The prevailing concept was that of "development-by-accumulation." In the traditional view (p. 1) "Scientific development *is* the piecemeal process by which ... items have been added, singly and in combination, to the emerging stockpile that constitutes scientific technique and knowledge." The task of science history was to chronicle these accumulations dating each new fact, law, or theory and identifying it with its "inventor," and identifying the errors and myths that plagued earlier scholars and stood in the way of scientific progress.

Kuhn discovered, however, that it became more and more difficult to fulfill the functions that such a view imposes upon science historians (pp. 2–3).

As chroniclers of an incremental process, they discover that additional research makes it harder, not easier, to answer questions like: When was oxygen discovered? ... Simultaneously, the same historians confront growing difficulties in distinguishing the scientific component of past observation and belief from what their predecessors had readily labeled "error" and "superstition." The more carefully they study, say, Aristotelian dynamics, phlogistic chemistry, or caloric thermodynamics, the more certain they feel that those once current views of nature were, as a whole, neither less scientific nor more the product of human idiosyncrasy than those current today. If these out-of-date beliefs are to be called myths, then myths can be produced by the same sorts of methods and held for the same sorts of reasons that now lead to scientific knowledge. If, on the other hand, they are to be called science, then science has included bodies of belief quite incompatible with the ones we hold today. Given these alternatives, the historian must choose the latter. Out-of-date theories are not in principle unscientific because they have been discarded. That choice, however, makes it difficult to see scientific development as a process of accretion. ... The result of all these doubts and difficulties is a historiographic revolution in the study of science, though one that is in its early stages.

It is clear from Kuhn's argument that the nature of that revolution is the development of a theory of the learning process of science that accords more adequately with the facts of its emergent history. Science history is not seen as a log book, but rather as record of the learning process. The historiographer has to ask the kinds of questions of history that will yield answers relevant to the description and analysis of a learning system.

Matson's (1964) book, *The Broken Image*, is another instance of an interpretive science history that suggests an emerging paradigm partially consistent with the concept of social learning. He describes how the machine system concept originating in classical Newtonian physics became highly restrictive when applied to biology, psychology, and social science. He goes on to point out that modern physics itself leads the way in introducing into scientific method the freedom necessary to build more appropriate biological and human sciences.

The major revisions he refers to are Heisenberg's "Principle of Uncertainty" and Bohr's "Principle of Complementarity." The former acknowledges the influence of the observer upon the observed and takes the first step toward making the scientist an endogenous component of the system he studies. In this book, we have extended Heisenberg's principle by making the social scientist not only an endogenous component of the social system he studies, but also one who seeks to change through evolutionary experimentation.

The principle of complementarity emerged out of the seeming incompatibility of the "wave" and "particle" theories of matter following the quantum revolution. Bohr solved the problem by pointing out that both theories were valid and essential to a complete explanation of matter. The two theories do not conflict because they cannot be applied simultaneously. They emerge out of two perspectives of the same reality. Through analogy this principle can be extended to assert the complementarity of and the need for both physiological and psychological theories in human science, both human system and social system theories in social science, and both steady state and social learning theories in social change.

Matson's development of some elements of this theme forms a good companion piece for this volume. It is well worth the attention of the reader.

Philosophy

Philosophy, too, is undergoing significant changes under the impact of the growing concern with learning systems. This is exhibited in several ways of which two are mentioned for illustrative purposes.

EVOLUTIONARY EPISTEMOLOGY

The concept of scientific progress as a kind of linear cumulation of tested ideas basically stems from the dominant epistemological view of nineteenth and twentieth century philosophy. It is essentially the "doctrine of sense data" which maintains that our sense impressions are the only thing we know directly and that all other knowledge is a logical construction. It emerged with Locke as a modern-day reaction to the idea of transcendental knowledge, or episteme, and it formed the basis of the logical positivism that has had such a strong hold upon the science community.

According to this attitude, for example, there would be no such thing as a paradigm shift in which one view of the research domain is replaced by another. If the initial view had any validity it would simply be enlarged by the extension of theory to accommodate more sense data. The data that existed before would not have changed, only the interpretation of them. Earlier theories would simply be special cases of later theories—unless they were false theories, in which case they would be categorically relegated to the graveyard of myth.

Such theories and the human effort that generated them are not accorded the status of science.

But the truth is that much of what we are beginning to learn about the performance of learning systems fails to support this kind of epistemological concept. According to Kuhn, for example, science history supports the view that:

> when paradigms change, the world itself changes with them. Led by a new paradigm, scientists adopt new instruments and look in new places. Even more important, during revolutions scientists see new and different things when looking with familiar instruments in places they have looked before. . . . paradigm changes do cause scientists to see the world of their research-engagement differently. Insofar as their only recourse to that world is through what they see and do, we may want to say that after a revolution scientists are responding to a different world. (p. 110)

> What occurs during a scientific revolution is not fully reducible to a reinterpretation of individual and stable data. . . . Rather than being an interpreter, the scientist that embraces a new paradigm is like a man wearing inverted lenses. Confronting the same constellation of objects as before and knowing that he does so, he nevertheless finds them transformed through and through in many of their details. . . . [The] interpretive enterprise can only articulate a paradigm, not correct it. (pp. 120–21)

The more we come to understand the operation of the learning process the more the epistemological paradigm that has dominated Western culture is being called into question and the more there seems to be emerging an evolutionary epistemology. It can be seen in Heidegger's (1949) philosophical approach to the concept of language and his distinction between the world of *stuff* and the world of *meaning*. It can be seen in Whorf's (1956) "principle of linguistic relativity." On the basis of linguistic analysis he hypothesizes that all higher levels of thinking are dependent upon language and that differences in the structure of languages one uses influence the manner in which one understands the environment. It can be seen in Popper's (1964) rejection of the model of passive induction as the basis for learning and his carefully elaborated "logic of scientific discovery." Campbell (forthcoming) gives credit to Popper for "recentering the epistemological problem." It can be seen in the exciting emergent work of Susanne Langer (1967)

269

in which she begins to treat both philosophically and empirically with the way "something called 'feelings' enters into the physical events that compose an animal organism" (p. 4). She maintains that feeling is a fundamental part of perception and intention and that our understanding of the learning process cannot be completed as long as we are bound to an epistemological paradigm that rejects its role and denies it a place in evidence. Bertalanffy's treatment of the mind-body problem (1964) strikes a similar note. He characterizes the weakness of the Cartesian dualism and advances the concept of the psychophysical organism. The emerging evolutionary epistemology can also be seen in Goudge's (1961) philosophical study of the theory of evolution. Finally, it can be seen in Lorenz's (1962) concept of gestalt perception which he suggests underlies the generalizing objectifying capacity of the interpretive process of the human mind.

In short, when we are faced with a learning process: "The central requirement becomes an epistemology capable of handling *expansions* of knowledge, *breakouts* from the limits of prior wisdom, *scientific discovery.* ... A focus upon the growth of knowledge, or acquisition of knowledge, makes it appropriate to include learning as well as perception as a knowledge process." (Campbell, forthcoming.) Philosophers, linguists, psychologists, science historians, etc. are all making a contribution to an emerging new epistemology that is not only consistent with the learning process but offers a paradigm that may help define the domain of normal science while improving our understanding of it.

Two special points ought to be made in calling these developments to the attention of the reader. First, it makes plain how truly basic and far-reaching are the consequences of our concern with learning systems. Second, it makes plain that the concepts advanced in this volume are based upon the presupposition of just such an evolutionary epistemology. The attempt has been made to make the contents of this effort generally consistent with what we know of it.

THE LARGER PHILOSOPHIC ENTERPRISE

The enterprise of philosophy is much broader than epistemology, of course. Within the domain of speculative philosophy it includes ontology, as well. Then there is the entire field of normative philosophy. This larger enterprise is also feeling the effect of change. Adler (1965), for example, points out that philosophy not only has a checkered past but is presently in some considerable disorder. The diagnosis offered is that philosophy has not conducted itself as a learning enterprise. Philosophical knowledge has not advanced in the way that scientific knowledge has because the conduct of philosophy has not met the conditions that are essential to a learning process. Included in these conditions is the necessity for philosophic knowledge to be of a form that is testable or subject to falsification and the necessity that philosophy be carried on as a public enterprise with cooperation in resolving disagreements. Langer (1967) makes a somewhat similar point in the introduction to her work. As another example, the modern ontology of Tillich (1951, 1957, 1963) places great emphasis on the fact that the "nature of being" only stands clearly revealed when interpreted in the light of the "process of becoming."

It would be a mistake, however, to leave the impression that these are developments in philosophy limited to the current period. Our debt to Charles Peirce (1893) has already been acknowledged in the prefatory quotation which opens this book. His disciple, John Dewey (1939), developed the broader philosophical concept of experimentalism to which Part III of this volume is obviously in debt. There are roots of modern process philosophy in Bergson (1911), Alexander (1920), Morgan (1923), Smuts (1926), and Sellars (1939). Many elements of the earlier process philosophies revealed in these works seem inadequate and inconsistent, but they form a clear line of emergence.

More recently Harris (1958, 1965) is adding to this line along with Tillich and others. Indeed, by 1966 the interest in process philosophy had reached the stage where a Society for the Study of Process Philosophies could be established.

The essential point offered here has two aspects: There has been emerging a process philosophy and an evolutionary epistemology. There is underway a reaching out for a weltanschauung, a world view or contextual philosophy which can survive the reality test imposed by the anti-entropic creative learning capacities that are demonstrated by physical, biological, and social systems open to the transfer of energy and information. This leads to a second effect. The philoso-

271

phizing about the enterprise of philosophy is, itself, going through a paradigm shift in which the ideas about the proper conduct of philosophy are being reevaluated. Central to that reevaluation appears to be the conviction that the practice of philosophy must itself be explicitly considered as a learning process and that traditional philosophic concepts need to be reinterpreted in light of the concept of the learning system.

Psychology

As with science history and philosophy, so psychology through its professional practice is entering into a new conceptual domain. Concern with the phenomenon of open learning systems is manifest in two aspects of psychological research: the processes of perception and cognition that are basic to human learning, and the concept of personality.

PERCEPTION AND COGNITION

Let us first look at Kuhn's (1964) interpretation. He calls attention (pp. 111–12) to the pioneering research of the Hanover Institute and the work of psychologists like Carr, Hastorf, and Bruner. He points out that experiments with glasses with inverted lenses and with anomalous playing cards demonstrate that human beings have the capacity to go through amazing transformations of vision. "Surveying the rich experimental literature from which these examples are drawn makes one suspect that something like a paradigm is prerequisite to perception itself. What a man sees depends both upon what he looks at and also upon what his previous visual-conceptual experience has taught him to see. In the absence of such training there can only be, in William James's phrase, 'a bloomin' buzzin' confusion.' "

This research appears to have developed some improvements in understanding the way perception is filtered through a cognitive grid established through prior experience. It has established that forgetting is an essential component in the process of learning. Its findings are among the anomalies that call into question traditional epistemology. Indeed, modern epistemology appears to be undergoing revision under the influence of the kind of reality testing that modern psychological research can bring to bear. A good place to see this process underway is in the work of Ittelson and Cantril (1954). They put forth a theory of perception based on modern psychological research that sees perception as "a transaction" taking place between concrete actors and concrete situations. Perception is viewed, in essence, as a dialogic process. This reinforces the significance of the analysis of communication as dialogue advanced in the text.

There is a striking isomorphism between this modern theory of perception and the concepts developed in the text. Ittelson and Cantril have this to say (p. 7): "This, then, is the central problem of perception: to study the degree of correspondence between the significances which we *externalize* and those which we *encounter* and to understand the process by which this correspondence is achieved. . . . it should be emphasized that the study of perceptual correspondence from this point of view also requires a consideration of the purposes of the perceiver." To make the parallel more plain one might say that people externalize certain aspects of their experience to create for themselves a world of things and people that forms a kind of operating context or boundary condition within which both perception and behavior take place. But the life encounters and perceptions are not only controlled by this externalized context, they, in turn, modify the context or boundary. Ittelson and Cantril put it this way: There is a ". . . process by which this correspondence [between the external boundary and life encounters] is achieved." This process seems clearly to be isomorphic with the process by means of which social goals and controls are made congruent in evolutionary experimentation.

THE CONCEPT OF PERSONALITY

As we saw in chapter V (pp. 178 ff.), until quite recently the area of research devoted to the concept of personality has been dominated by two conceptual models.

On the one hand are those represented by their proponents as the "truly scientific" movements. These include environmentalism, stimulus-response psychology, mathematical models, and psychological geneticism. In general, these branches of psychology look upon the mind

273

as an open vessel that is filled during the course of ontogeny by a stimulus-response mechanism under the motivation of tension reduction. They hold that all behavior tends toward equilibrium or the elimination of an excited state. They maintain that all tension and all striving have their origin in the disturbance of organic equilibrium. They picture the organism as an essentially reactive system responding to external stimuli.

In contrast is the psychoanalytic movement which maintains that adults seek patterns of behavior and personal relationships that were learned to be satisfying in childhood and probably arose out of childhood solutions to some problem of personal anxiety. Thus, patterns of behavior gratify needs laid down during the ontogeny of the personality.

While psychoanalytic theories appear to be quite different from stimulus response theories in many respects, both display an important common attribute. Both conceive of the human personality as a homeostatic system addressed to tension reductions. Bertalanffy (1962a, p. 14) makes the following observation: "According to Freud it is the supreme tendency of the organism to get rid of tensions and drives and come to rest in a state of equilibrium governed by the 'principle of stability' which Freud borrowed from the German philosopher, Fechner."

It is only recently in the work of Allport (1955), Buhler (1959), Maslow (1968) and others, that a more open psychological theory has begun to emerge. Buhler underscores the change in the following way (p. 576): "In the fundamental psychoanalytic model, there is only one basic tendency, that is toward *need gratification* or *tension reduction....* Present day biological theories emphasize the 'spontaneity' of the organism's activity which is due to its built-in energy."

The reader will recall that this theme was developed more fully in chapter V, where Allport's distinction between the individual's growth motives and deficit motives was highlighted. The deficit motives were identified with the tension reducing motives of biological maintenance. The growth motives were identified with the tension-seeking motives of the spontaneous creative human being.

It is fascinating to observe the degree to which another discipline working on different problems with different methods is moving into the same domain of learning systems, and how, in the process, it has generated so many concepts with such a familiar ring. Once again we find behavior differentiated into static state and developmental modes. The "deficit motives" or "tendency to maintenance" of human personality appear to parallel the concepts of adaptive specialization and equilibrating adjustments in biological and social evolution. Both are stability seeking. Indeed, personality maintenance and biological maintenance seem to enter as basic motivations into the generation of social maintenance.

In contrast, the "growth motives" of the individual seem analogous to adaptive generalization in biological systems and evolutionary experimentation in social systems. Many of the ideas of normal personality development sound strangely like the characteristics of adaptive generalization discussed in chapter II. The growing personality improves its "integration" or "organization." It has a progressively more complex idea content and gains in versatility and plasticity. It becomes progressively more cerebralized and remains "open" to further development.

Additional parallels with evolutionary experimentation in social system learning are apparent, particularly in the writings of the developmental psychologists like Piaget. (See Flavell, 1963.) According to this view, the cognitive development of the human personality goes through certain observable developmental periods when the cognitive organization experiences pronounced shifts. In between are relatively stable developmental plateaus in which personal learning takes place under the guidance of (and in a way generally consistent with) the prevailing cognitive organization. There come times, however, when the life encounters cannot be adequately assimilated in terms of the prevailing organization. Periods of personal crisis and personal reintegration then take place. How familiar this all sounds! It immediately evokes the image of the normal problem solving and boundary shifts that characterize social system learning. In fact, there seem to be concept parallels here that might fruitfully be pursued much further.

Cybernetics and General Systems Theory

In recent years an embryonic discipline has emerged devoted to the study of the control processes in machine systems and life systems. It was introduced to the scientific community about twenty years ago with the arrival of Norbert Wiener's celebrated book (1948). Although the roots of such a study take a number of forms and go back in time, recent years have provided a new name and visibility along with a dramatically expanded focus of effort. As yet, there is no agreement among the practitioners concerning terminology, but the study of control systems is tending to divide into cybernetics and general systems theory, the former being a special case of the latter.

Cybernetics found its origin primarily in the study of the control of machine systems.[1] Because of this orientation the early work was largely absorbed with defining and studying homeostatic processes. It concerned itself with systems that are deviation counteracting—that go to equilibrium—and with the deterministic controls that operate to eliminate the nonequilibratory states of the system.

As the studies began to deal with life systems as well, greater emphasis came to be placed upon deviation amplifying processes (positive feedback) along with deviation counteracting processes (negative feedback). For example, in the study of control processes in living organisms, Stanley-Jones and Stanley-Jones (1960) recognize the need to see the process as a combination of deviation amplifying and deviation counteracting controls. Such hybrid control systems are also of interest in machine systems. Jet engines and atomic piles, for example, are deliberately designed as runaway systems to get the greatest possible output, but are kept in leash by negative feedback that controls the rate of output.

It is interesting to see that this emerging discipline is concerned with the theory of system boundaries and controls. Its significance for the study of the role of organization, which has absorbed so much of our attention in this volume, is apparent. But at this early level it has nothing to say of relevance to the process of social learning, because it does not deal with the processes that have the capacity to reformulate their own controls. It is not surprising, therefore, to see some of the more recent work moving in this direction.

One thrust of this effort tries to move into the domain of the learning system by extending the concept of the machine system. We find

---

[1] For the reader new to the field there are two basic introductions: Cherry (1957) and Ashby (1964).

the discipline considering such topics as the "machine which invents" or "the principle of the self-organizing system." Ashby's (1962) discussion of this approach was extensively cited in chapter I (pp. 27 ff.). It was pointed out that the concept of an object machine under the control of another machine which has the capacity to reprogram it does not reveal a way out of the impasse of deterministic systems because it leads to an infinite regress of higher order determinisms.

There are two other reasons for the inadequacy of this approach to a solution. First, creative learning systems have to be fully open systems. As Bertalanffy (1962a) points out, it is not sufficient for a machine system to be open to information as Ashby proposes. It must also be open to entropy transfer if it is to have the capacity to self-generate a "progressive differentiation, evolving from lower to higher states of complexity." But if it has this character, one can no longer stretch the machine system concept to accommodate it. Second, while a machine with input in Ashby's meaning may make a contribution to behavioral modification within the contexts of the goals and processes made explicit by a given paradigm or conceptual framework, it has no capacity for negotiating a paradigm shift. It cannot reorganize anomalies into a new paradigm. At this stage of our understanding this kind of creative capacity appears to be clearly reserved to the human mind.

As cybernetics-oriented investigation has evolved, the scholars concerned with these issues have refused to stop at this dividing line. A few hardy souls are pushing beyond into the domain of systems open to both information and entropy transfer and capable of acting as true learning systems. As this has taken place the discipline seems to be changing in two ways: (1) It is coming to be characterized more and more by its practitioners as a general systems theory. (2) The field of investigation is becoming less dominated by mathematicians emphasizing purely deductive systems and has attracted a wide range of substantive specialists in the social and life sciences who are adding a decided empirico-intuitive character to the enterprise. The work of Bertalanffy (1962a, b, 1968) probably forms the best introduction to this broader orientation to the study of systems.[2]

[2] Anyone interested in this developing literature would also do well to examine all of the volumes of the *General Systems Yearbook* sponsored by the recently established Society for General Systems Research.

Bertalanffy (1968, pp. vii, viii) emphasizes the need for this broader orientation in the following words: "The student in 'systems science' receives a technical training which makes systems theory—originally intended to overcome current over-specialization—into another of hundreds of academic specializations. Moreover, systems science, centered in computer technology, automation and systems engineering, appears to make the systems idea another—and indeed the ultimate— technique to make man and society even more into [a] 'mega-machine.' ... What may be obscured in these developments—important as they are—is the fact that systems theory is a broad view that far transcends technological problems and demands, a reorientation that has become necessary in science in general and in the gamut of disciplines from physics and biology to the behavioral and social sciences and to philosophy."

The work we have undertaken in the present volume appears to be in this broader tradition of general systems research.

Decision Theory and Organization Theory

It is a bit difficult to justify the distinction between decision theory and organization theory that grows out of common practice. One would be hard put to it, for example, to define the way in which decision theory differs from organization theory or how they both differ from the system theory discussed in the last section, yet they all are absorbed with the study of the control processes characteristic of systematic behavior. Because systems research emerged from different disciplines and problem contexts, various aspects of the study of the control process have come to use different labels. This suggests the need for a more comprehensive theory of general systems that can progressively bring all systems research under a common umbrella.

The intention here is simply to identify certain instances that illustrate the movement of the research disciplines toward something that might be characterized as a social learning paradigm. Several developments of this kind are taking place under the traditional rubrics of decision and organization theory.

First, consider the changes evolving in the field of decision theory.

Decision theory grew quite naturally out of the classical economic assumptions of rational, competitive, optimizing behavior. It is addressed primarily to the decision problems of the business enterprise operating within the general equilibrium market economy. The classical economic theory defines the decision rules where information about the states of the system is certain. Faced with uncertainty in reality, decision theory has moved to develop decision rules that lead to rational solution in the face of uncertainty. It has done so by modifying classical statistical mechanics with the aid of the Bayesian theorem to produce a "practical" way of choosing the "best" act. This is done by assigning "values" to consequences and probabilities to parametric events and then selecting the act with the highest expected value. The "decision matrix" is structured to accommodate "subjective probabilities" as well as the objective probabilities of classical statistics. The method has been applied to a range of business decision problems—notably those dealing with inventories, production scheduling, and quality control. Game theory is another manifestation of a concern with the mechanics of choice.

It is apparent, however, that the useful applications of decision theory have been restricted to quite well-structured decision problems. A period of reexamination and reaching out seems to be underway. Evidence of this is offered by a recent study, "Planning on Uncertainty," by Ruth Mack (forthcoming).

Dr. Mack points out that as one moves out of the domain of well-structured problems one's orientation to the decision problem of necessity becomes altered. She says, "on the one hand, efficiency optimizing in the conventional economic sense is beset by ambiguities of measurement and even meaning. On the other hand, the creative aspects of decision become heavily underscored—the aspects that continue to increase the utility of outcomes by wiser deliberative methods, improving motivation and providing scope for innovation." She emphasizes that the new orientation requires a shift from emphasis upon the "decision proper" to a "decision process." "Decision itself . . . is only one incident in a life history. The history starts with the events that bring some problem into focus and ends with the last efforts to correct and carry through the actions undertaken."

Dr. Mack progressively puts together a highly suggestive "checklist" of decision rules that may prove useful in guiding decision processes in unstructured and open-ended decision contexts. She calls this "quasi-optimizing to emphasize the fact that there is inherent lack of precision in all aspects of the procedure, at the same time that effort looks toward considerations commonly attributed to optimizing." She sets as the proper objective "progress toward rather than the achievement of goals. . . . proper goals are always unachievable. They are an arrow and not a point."

It should be clear from these few references that Dr. Mack is struggling to cope with decision rules appropriate to a "learning" decision maker acting in an operational environment that is characterized by a process of social learning. It is a line of investigation that deserves careful attention and continued development.

Like decision theory, organization theory grew out of an early concern with the problems of industrial management. But the focus was slightly different: Instead of centering on the process of decision making, the primary emphasis was on the structure of the controls that implement the decision. In its early phases attention was absorbed by concepts of formal hierarchical organization designed to serve the management of essentially homeostatic systems. By making extremely simple assumptions about human motivation and response, human components were treated essentially as machine system components.

Later, much more attention was devoted to the way in which human participation in social organization is controlled and the way in which social goals and individual goals can be reconciled through appropriate modes of social organization. The work of Likert (1967) has already been cited. The earlier works of Argyris (1957, 1960) should also be noted. While in this phase the emphasis is on formal organization and the objectives of promoting system maintenance and productivity, a movement in the direction of learning system concepts is perceptible: the role of human creativity is brought into social organization by the back door.

Today, organization theory is experiencing a new development. Writers like Pelz and Andrews (1966) and Bennis and Slater (1968) are becoming concerned, not with organization for social maintenance or productivity narrowly conceived, but with the organization of social change. This was discussed rather fully in chapter VI. Here it is sufficient to remind ourselves that the field is only beginning to grapple with those issues that are of interest to social learning.

Finally, special mention should be made of a work that combines some of the features of both decision theory and organization theory. Sackman's *Computers, System Science, and Evolving Society* (1967) illustrates remarkably well the emergent character of the social learning paradigm and is one of two works [3] that comes closest in spirit to the intent of the present volume.

Sackman writes out of a background of experience in the description of realtime man-machine control systems based on the new computer technology. Development of the SAGE air defense system led him to the realization that such large-scale complex control systems are not engineered in the usual sense. They are the product of a continuous process of what Sackman terms evolutionary experimentation. This orientation becomes especially important as he projects what might happen if the lessons of this experience were extended to the improvement of still larger and more general man-machine control systems.

The evolution of computers has broadly progressed from mathematical computation to information processing and finally to realtime control. The evolution of science has likewise progressed from the pursuit and collection of knowledge to the mastery of the natural environment and control of events. In both cases information is eventually followed by control.

The marriage of information to control shatters whatever pretense may have existed for watertight domains between science and the humanities. The application of scientific information toward social control places human values and scientific endeavor in the same boat, sharing the same destiny and ultimately a common fate. (p. 23)

A man-machine digital system is an evolving organization of people, computers, and other equipment, including associated communication and support systems, and their integrated operation to regulate and control selected environmental events to achieve system objectives. (p. 42)

A basic tenet of the proposed approach to scientific system development is that each man-machine digital system be conducted as a scientific enterprise, experimentally investigating its own behavior for self-corrective adaptation to changing system conditions. This implies that each system should investi-

[3] The second work, by Etzioni (1968), is dealt with on p. 294.

281

gate its own realtime behavior to effect better understanding and more effective control over system events. We cannot normally expect any currently established science to look closely and in exacting detail at the real time activities of any existing system, especially a highly unique and complex system. But we can expect those responsible for and involved in the object system to have the motivation, interest, and first-hand intimate knowledge necessary to examine the real time activities of their system closely and, with the aid of the central computing facility, to become increasingly adept in understanding and controlling the course of system events. The transition to scientific system development is effected by the application of experimental method toward the regulation and control of real time system events. . . . science exists wherever experimental method is cultivated. (p. 225)

Can system development be conducted as a scientific enterprise? There is a conceptual analogy between system development and experimental method . . . . The gist of the analogy is that system development may be conceived as an evolving set of hypotheses relating system design to system performance, subject to experimental verification throughout the life cycle of the object system. (p. 206)

The central point is almost disarmingly simple: operational specifications define system design and make claims for system performance; treating them as working hypotheses serves notice to everyone concerned that all such claims are continually subject to impartial testing to separate fact from fancy. This is not a plea for transforming system development into pedantic laboratory exercises. It is a hardheaded demand that each system deliver the goods as promised.

. . . each system becomes the object of its own experimental investigation. Hypotheses and tests are generated within the system to be tested by the system. This implies that managers, users, operational designers, programmers, and operators are integral parts of the evolutionary testing process and, at various times, are either experimenters, or subjects, or both, testing themselves and evaluating their own performances and that of others. (p. 207)

Economic Development

Since this volume was born out of a professional concern with the problems of economic development, we have already observed some of the shifts of concept taking place in this field. Chapter I was devoted in part to a short characterization of some traditional model concepts. Here, this discussion is extended to indicate how the economic literature is handling these ideas and how the emphasis has been changing.

Simple Growth. As is true of all these concepts, the simple growth concept of economic change has both a passive and an active form. The passive form is represented by the Malthusian theory of deterministic increase in population up to a subsistence margin imposed by resource constraints. The active form represents the beginning of the economist's absorption with the role of capital. It is recognized that the dismal subsistence margin can be avoided if per capita productivity is enhanced by expanding the levels of behavior-amplifying human artifacts (material capital) and by controlling the level of population. It yields not a passive historical determinism, but an active system-engineering determinism. In the end, of course, it yields only a displaced Malthusianism. If one is concerned with the levels of capital relative to the other productive factors, the increase in the relative level of the capital stock would generate an optimum level of productivity beyond which further increases in the size of capital stocks would force the marginal productivity of capital below that of other factors and force a decline in productivity. The passive Malthusian optimum yields a maximized population; the active capital formation model yields a maximized per capita productivity. But the implacable diminishing margin operates in each case.

The System Open to Resource Exchange. The implacable diminishing margin, it turns out, can be made less implacable if the system is open to resource exchange. Hence, economic concepts become enlarged to include this metaphor. This yields a focus upon the importance of the "economic base" and interregional-international trade in fostering growth. Under this concept the relatively advanced systems could displace the implacable margin by expanding their natural resource base through trade (or war). The less productive systems could displace the implacable margin more easily by expanding capital stocks through trade. This might require some allometric adjustments in the internal mix of activities as well as the overall levels of capital, but this is achieved by balancing the "marginal social productivities" of the dif-

283

ferent forms. If this proves difficult where market institutions are underdeveloped and prices are unreliable as a guide to real product, one may have to work with "shadow prices" or "accounting prices" or fall back upon such procedures as crude capital output ratios, but the basic rationale is always the same.

This metaphor has run into difficulty, of course, because it focuses upon growth rather than development—with the scale and proportion of factor inputs rather than innovative changes in modes of behavior. It is still concerned with the displacement of the implacable margin rather than the transformation of its character. Accordingly, in time the economic literature came to adopt the metaphor of a system open to information exchange.

The System Open to Information Exchange. The concept that the system can be open to the transfer of new ideas has not been as well developed or consolidated in the economic literature as that dealing with the exchange of resources and capital, but the movement into this domain is clearly recognizable.

In its most developed form it accepts the notion that the capital transfers to underdeveloped systems can take the form of capital-embodying behavioral ideals novel to the importing system. Improving the behavioral capacity of that system is seen to rest not merely on the level of the capital stock, but on its form as well. This yields the notion that the behavior of the system can be reprogrammed through the exchange of forms of physical capital. There emerges a simple learning model, but it is essentially the deterministic model of programmed learning.[4]

Each of these three successively more general growth and development models reinforces the economist's propensity to view change as a process of movement from an initial to a terminal state. They are exercises in comparative statics. The program for development is derived by subtracting the present state of a social system from its desired state. The "differences" that receive most attention from economists have been differences in the sizes of the capital stock relative to the population and differences in the forms of physical and human capital.

[4] See chapter I, fn. 3.

Encouraging economic growth and development is essentially a problem in capital formation and population control.

Emerging in the literature is a realization that the models on which economists have depended are not adequate. Some of the ways in which the problem is being attacked are described below.

Arrow's "Learning by Doing." The fact that even the traditional heartland of economic theory is reaching out to incorporate learning concepts is illustrated by Arrow (1962). Writing in the Harrod-Domar-Kaldor tradition of the theory of economic growth, he emphasizes that increases in per capita income cannot be explained by increases in the capital-labor ratio. He suggests that the important missing factor in the traditional formulations is the fact that knowledge is growing in time. Drawing upon the concept of the learning curve from psychological learning theory, he emphasizes the importance of the role of experience in increasing economic productivity. He suggests that cumulative gross investment as an index of experience may be a modification of the traditional formulations of aggregate growth theory.

In terms of the perspective of this volume this is obviously a very limited application of the learning concept. It is employed as a correction factor within the context of the traditional economic paradigm, rather than a new context reordering our interpretation of the process. However, it is just such clues as this that indicate the extent to which the practice of social science under the control of the old metaphor is experiencing a strain resulting in a reaching out for new concepts.

Stage Theories of Economic Development. One of the ways that the traditional concept has become enlarged in the literature of economic development is through the presentation of the process of new capital formation as a series of stages. The simple initial-terminal state image yields to the image of a series of successive states.

The tendency here has been to turn to economic history to identify the stages characteristic of developing systems. For example, both the Clarkian (Clark, 1940) and Rostovian (Rostow, 1960) stage theories are generalizations derived from very large-scale features of economic history. Clark represents the historical progression to higher levels of productivity as an emergence through "primary, secondary, and tertiary" states representing a movement from agriculture and other natural resource-oriented activities, through processing activities like manufacturing, to an emphasis upon human services. Rostow sees societies as moving through five stages consisting of the traditional society, the preconditions for "take-off," the "take-off" itself, the drive to maturity, terminating with the age of mass consumption.

Basically, the Rostovian scheme is an extension of the traditional concepts of capital formation already outlined. He is similarly concerned with the displacement of a traditional society to a level of modern mass consumption. However, he realizes that the transition from one to the other is the result of a set of sequences that must be specified for the capital formation model to prove useful to the planning process, hence his concern with outlining stages. In addition to the basic concepts of capital formation, he incorporates a number of other concepts including the Clarkian industrial stage sequences, the economic base multiplier, and a broader concept of psychological motivation. He introduces the notion that these are critical thresholds in the rate of capital accumulation. The result is a historical statement of broad scope and imagination containing many highly suggestive and enlightening ideas.

However, we have learned (in chapter IV) that historical statements have a limited utility for positive prediction. The implication of the stage theory concept is that it provides a template for planning— particularly for planning the stages of new capital formation for societies following in the footsteps of their more advanced predecessors. It has tended to exhibit two limitations in this regard.

The first stems from the assumption that the historical stage sequences offer an adequate model of the sequences necessary in every case. In fact, the emergent line of every system need not follow this sequence.[5] This has been recognized by Gerschenkron (1962). He is

---

[5] The emergent line of every social subsystem need not follow the same sequence at all. This was pointed out some years ago by the author (see Perloff et al., 1960, chap. 4) when it was noted that under some circumstances a region may be opened up under the influence of an expanding and transforming urban megasystem in a sequence that moves from tertiary activities back to certain associated secondary activities. etc.

sensitive to the historicist error that resides in the assumption that all economies are supposed to pass through the same unilinear stages as they move along the road to economic progress. In his study of European industrialization, for example, he has classified countries according to degrees of backwardness and concludes that the ". . . differences in point of departure were of crucial significance for the nature of subsequent development." He seeks through his work to ameliorate this first limitation of the stage theory concept.

There is a second limitation, however, that none of the current stage theory concepts seem to escape. They do not deal directly with the processes by means of which new behavioral ideas come into existence and become embodied in actual behavior through social reorganization. They do not come to grips with the fundamental character of economic development—social learning. The fact that this difficulty is not altogether overlooked by economists can be seen in the writings of the new technologists.

The New Technologists. The model of the system open to information exchange has been subject to a second form of modification in the economic literature. There is a growing tendency to see development as something more than the transfer of developed forms of capital to underdeveloped systems.

There is an increasing awareness that new behavior modes embodied in real capital do not transfer readily to areas where the existing stocks of natural and human resources along with the traditional modes of social organization are quite different. One consequence of this realization is the emphasis given to technological adaptation by Gunnar Myrdal. In a recent address he pointed out that an essential ingredient in the development of the underdeveloped regions of the world is an endogenous emphasis upon technological research aimed at adapting more advanced forms of capital to the specific condition and needs of the developing region.

A more general example of this new awareness is found in a book devoted to technology and economic growth by Nelson, Peck, and Kalachek (1967).[6] In the opening paragraph they go directly to the

---

[6] This book also serves as an excellent cross-reference to other publications growing out of this recent and still limited professional interest.

heart of the inadequacy of conventional concepts. "In most of the literature on economic growth, technological advance tended to float in the air as a factor which increased the productivity of capital and labor, its contribution being estimated sometimes by the residual in the growth rate after allowance for growth explained by other factors, sometimes by a time trend in productivity." They go on to identify technological advance explicitly with the creation and embodiment of new ideas.

This is definitely a study addressed to a better understanding of some components of the process of social learning, although the learning concept is not explicitly employed. It is a useful study and an example of a reorientation in research interest that one hopes will become more common among economists. Because it is an exploratory and pioneering effort in a new domain, it is no criticism of its contribution to point out that its scope is limited—designedly so. First, the authors focus on technology, restricting their attention to that component of social learning which constitutes the generation of behavioral designs—with emphasis restricted further to the modification of machine system behavior devoted to the material and energy throughputs of society. Within this restricted frame they see the rate of technological progress as primarily a function of the volume of resources applied to research and development which is, in turn, modified by such factors as (a) the magnitude of the change in performance being sought, (b) the complexity of the system, and (c) the stock of knowledge available for recombination into new behavioral design.

While the scope of this effort is limited, it constitutes an important piece of work in a new domain for economists. Research packages like this will cumulatively add to our understanding of the process of social learning.

At a still more general level we have the writing of Schon (1963, 1967). From an initial focus upon the impact of invention and innovation upon American economic development, Schon (1967) moves to a consideration of the process by means of which invention and technological change take place. He criticizes the "rational view of invention" which sees invention "as the conversion of knowledge to technology" by an orderly, goal-directed, intellectual process. He characterizes it instead (pp. 8, 11) as a "complex process in which goals are discovered, determined and modified along the way.... A continuous social process without clear beginning or end." In a similar way he rejects the rational view of innovation, indicating that planning for innovation often takes the form of "leaps of decision" that cannot be supported by available information. In the process of developing his ideas he formulated the idea of the "displacement of concepts" in the interest of capturing the essence of the developmental process. This "displacement of concepts," it turns out, is quite similar to the idea of Kuhn's paradigm shift that has been developed so intensively in this volume. It is clear that Schon's interest in technology has led him into a domain of concept that can only be characterized as a social learning model.

The Role of Human Capital. There is still a third way in which the model of the system open to information exchange has become modified in the economic literature. The earlier focus upon physical capital has been extended because of an increasing realization of the importance of human capital.

At first the focus was dominated by a concern with the fact that new forms of physical capital required new human skills to complete the installation of the necessary man-machine system. The problem of increasing the stocks of established human skills and developing new ones to match novel forms of physical capital began to absorb the attention of economists. The writing of Schultz (1963), Denison (1962, 1967), and Bowman and Anderson (1965) can be cited as examples of this interest. The focus, however, was upon the influence of increases in the stocks of human capital and the devices by means of which they can be augmented. There is little disposition to go beyond the notion of programmed learning as a means of embodying known skills into human components of social systems. There is a tendency to treat human beings as social artifacts.

A more general focus explicitly takes into account the open creative contribution that human beings make to the process of economic development. This is reflected in the writing of McClelland (1961) and Hagen (1962).

McClelland is a psychologist who has become interested in the problems of economic development. He places great emphasis upon the

role of psychological motivation in development. He points out that
people's psychological need for achievement may differ both as between
individuals and between social groups. He reasons that high rates of
development are associated with high levels of "need achievement." He
displays a great deal of ingenuity in developing measures of the differences in need achievement between different historical groups at
different historical periods, and relates these fluctuations in the levels
of need achievement to the growth rates of the social group. He finds
high levels of correspondence, with growth and development lagging
behind the fluctuations in need achievement.

McClelland's work establishes the statistical significance of these
creative psychological factors. Hagen (an economist interested in psychological concepts) attempts to move beyond this to the development
of a hypothesis about the process of social change based upon psychological theory. It is, in effect, a hypothesis about an evolutionary sequence in which economic development and personality development
are dynamically interrelated. It is based upon the psychoanalytic
theory of personality development which maintains that the basic personality traits are formed in childhood and are primarily a product
of the parent-child relationship.

Hagen points out that at the extremes of a spectrum two types of
societies can be identified—traditional and innovating (read steady state
and learning). Associated with each is a characteristic personality
type—authoritarian or innovative. The presence of an authoritative
culture constitutes an almost overriding obstacle to the behavioral
transformations essential to development. Consequently, the "transition to economic growth" depends upon some sort of shock to the
authoritarian personality. Such a shock takes the form of a "... perception on the part of the members of some social group that their
purposes and values in life are not respected by groups in the society
whom they respect and whose esteem they value" (p. 185). The occurrence of these shocks is not systematically analyzed and they are considered to be in the nature of historical accidents, although Hagen
identifies several forms in which such shocks have occurred.

Once traditional society has been exposed to such a shock, there
follows a series of adaptations in personality so that each successive
generation forms a different personality type as the new personality

type of the parent impinges upon the child during its formative years to breed a still different personality type in the succeeding generation. Hagen (p. 193) explains it this way: "The first effect seems to be the appearance after one or more generations of a type of personality which I shall term 'retreatist'; or, if the social tensions are more severe, a type which I shall term 'ritualist.' Out of ritualism, in turn, retreatism may develop. Later, retreatist personality may give way to innovational personality to be termed 'reformist.' " Development takes place under the influence of the innovative personality. Hagen presents a number of historical case studies where, according to his interpretation, the facts of social development are consistent with his hypothesis.

In short, a social system appears to require as a precondition to its development an adequate number of individuals who are satisfying their "growth motives" through creative action. Both of these studies emphasize quite a different aspect of development from that which grows out of the economist's more traditional concern with capital formation. They are genuinely moving into the domain of social learning.

However, while this work represents an important effort to extend traditional concepts of economic development by incorporating one factor important to social learning, it is still very incomplete. It is also curiously limited when evaluated in terms of its objective of exploiting psychological theory. Hagen is still locked into the psychoanalytic model of the personality and is a victim of the same defects that are prompting the major changes in concept currently taking place in psychology.

As a consequence, for Hagen the development of a personality in which innovation comes into play is largely an accident of the environment. In particular, it is the product of the child's reaction to parents during its formative stage. The innovative personality is really seen as a means for reducing tension generated in the parental relationship. This tension generates a "need for achievement" that may be discharged through innovative action. By clinging to the psychoanalytic theory, creativity is transformed into a deficit motive. It is also transformed into a reactive phenomenon.

In order to explain the generation of sufficient innovative personalities to support social change, Hagen has to have recourse to a model of change strikingly similar to phylogenesis in its mode of action. It begins with a shock administered by environmental change upon the authoritarian personality of traditional societies. This yields personality conflicts within the home that foster in offspring various ways of reducing these tensions that may, in the course of several generations and intermediate states, lead to the establishment of an innovative personality that resolves tension through creativity. When the population acquires a sufficient number of these, the transition from a traditional to an innovative society is achieved.

There are two difficulties here. First, Hagen established no better record than the biologists in dealing with the question of the "fitness of the environment" to support this transformation in the "personality pool" over several generations through the operation of the process of reproduction. The occurrence of an environment that will stimulate such a change is considered a historical accident. The environmental attributes that form this precondition are not analyzed and the conditions under which a favorable sequence of environmental change and personality adaptations might be sustained are not considered. As we observe in the case of biological evolution, the phylogenetic model does not describe a complete learning system. Hagen, similarly, leaves some of the most important considerations unexplained.

Second, without necessarily denying that some forms of creativity can sometimes be motivated by the need to reduce tension, the emerging theories of personality would maintain that creativity can be actively as well as reactively motivated and that these patterns are not formed exclusively in the relationship of child to parent. Hagen's thesis would doom social change to the classic scenario of phylogenesis and the wasteful time-consuming process of discontinuous transformation of population groups. It denies the possibility that creativity and growth motives may be positively rather than negatively stimulated in different ways and during different phases of the ontogenetic development of the individual. It does not accommodate the possibility that behavior-changing behavior or social learning may, with the aid of the emerging work in psychology, learn to generate creative personalities through more direct methods. And it ignores the possibility that social learning may be already operating in a less consciously understood and organized way to generate such personalities.

Development as a Cognitive Process. Among the writings of those who are searching for a new metaphor or paradigm to guide the interpretation of the developmental process, Solo (1967) presents a conception of social change that coincides with many of the ideas presented in this volume.

Solo sees economic revolution as a revolution in cognition. He states that "the organization of production, the organization of the transformation, and also the organization of whatever creates the goals for transformations . . . are different processes." He emphasizes a need for a power to transform that is separate from operations. He emphasizes knowledge. ". . . economic revolutions appear as the successive extension of the role of cognition." (p. 107)

Boulding (1961) suggests a similar metaphor when he advances his concept of "the image." The basic proposition of this work is "that behavior depends upon the image" that motivates it and controls it. He emphasizes that "the image is built up as a result of all past experience of the possessor of the image. It has a historical developmental dimension. . . . The subjective knowledge structure or image of any individual or organization consists not only of images of 'fact' but also images of 'value.' " (pp. 6, 11)

This metaphor of the cognitive process is a direct engagement with aspects of the process of social learning. Both studies are well worth the reader's examination, but neither of them offers a comprehensive view of the process of evolutionary experimentation at work to transform the behavior of social systems.

The World of Kenneth Boulding. In the above context, one cannot leave the work of Kenneth Boulding (1953, 1961, 1964, 1966, 1968) with so casual a reference. More than anyone else in the field of economics Boulding has made repeated forays into the domain of social learning. It is difficult to overstress the imaginative, provocative, and unfettered character of the mind of this man. He has scouted a great deal of territory and has consistently called upon social scientists to stretch beyond their traditional domains. There may be few topics or major conclusions presented in the present volume that Boulding has not foreshadowed.

That Boulding ranges over much of the same intellectual landscape can be seen by the fact that he is an early champion of general systems theory and frequently employs the analogies of embryological or ontogenetic process, biological ecology, and the large-scale features of phylogenesis. He has consistently sought to reach beyond the traditional domain of economics to acquire a more comprehensive view of the social process.

At the same time, Boulding's work does not add up to a comprehensive, integrated treatment of the theme. His writing is chiefly in the form of essays and articles. In his utilization of the various biological analogies he has never indicated clearly the differences between them or the degree to which they may legitimately be taken as analogues for the social process. Those who have been stimulated by his imagination and insight hope that he may yet weave these creative strands of thought into a whole cloth.

Sociology and Social Psychology

Along with the work of Sackman, the recent writing of Etzioni (1968) comes closest to presenting a comprehensive view of the process of social learning. While the image of the learning process is not invoked and quite a different language of societal analysis is employed, most of the basic elements are there. Indeed, this work anticipates so much of the intent of this volume that it would have received extensive citation throughout the text had it not been published after the author had completed his manuscript. Here, at the end of this book, therefore, we take the opportunity to acknowledge an impressive creative effort to encompass the domain of social learning.

The degree to which there is a congruence of ideas can be illustrated with a few words from the Etzioni book:

A central characteristic of the modern period has been continued increase in the efficacy of the technology of production which poses a growing challenge to the primacy of the values these means are supposed to serve. The postmodern period, the onset of which may be set at 1945, will witness either a greater threat to the status of these values by the surging technologies or a reassertion of their normative priority. Which alternative prevails will determine whether society is to be the servant or the master of the instruments it creates. The active society, one that is master of itself, is an option the postmodern period opens. An exploration of the conditions under which this option might be exercised is the subject of this endeavor. (p. viii)

A societal unit has *transformability* if it also is able to set—in response to external challenges, in anticipation of them, or as a result of internal developments—a new self-image which includes a new kind and level of homeostasis and ultra-stability, and is able to change its parts and their combinations as well as its boundaries to create a new unit. This is not a higher-order ultra-stability but an ability to design and move toward a *new* system *even if the old one has not become unstable.* It is an ability not only to generate adaptive changes or to restore new stability to an old unit, but also to bring about a new pattern. (p. 121)

Actors whose decision-making is based on a mixed-scanning strategy differentiate contextuating (or fundamental) decisions from bit (or item) decisions. Contextuating decisions are made through an exploration of the main alternatives seen by the actor in view of his conception of his goals, but details and specifications are omitted so that overviews are feasible. Bit-decisions are made "incrementally" but within the contexts set by fundamental decisions (and reviews). Thus, each of the two elements in the mixed-scanning strategy helps to neutralize the shortcoming of the other. (p. 283)

An active approach to societal decision-making requires two sets of mechanisms: (a) a high-order, fundamental policy-making process which sets basic directions and (b) an incremental process which prepares for fundamental decisions and revises them after they have been reached. When rapid changes in the environment, in the societal unit, and in the step-structure nature of the problem, or prolonged mistaken treatment, lead to increasing difficulties, the higher order, fundamental review process must become operative. (p. 290) [Recall the paradigm shift and normal problem solving!]

An intimate link exists between the nature of man and the nature of the social grouping which commands his prime loyalty and which is the center of his public life....

In the process of societal activation, not only do more people gain a share in the society, hereby reconstituting its structure, but the members themselves are also transformed; they advance along with the society they are changing. ... There is a need for dynamic interchange between personal self-realization and societal activation. (pp. 11, 15) [We encounter here the interface between individual and social group learning.]

The main questions for the transformation toward an active society are whether or not societies can mobilize themselves and their member collectiv-

295

ities to high, crisis-like if not higher, levels in noncrisis situations, and whether or not they can generate power for internal self-transformations instead of exerting their wills on other societies. Further, can this level and kind of mobilization be attained without generating so many counter-currents and so much alienation that the consensual base of society and values related to it will be undermined as the realization of the values expressed in the goals advanced is enhanced? In short, is a "permanent revolution," a continual and authentic social-movement society, possible (p. 399)? [Another expression of the fourth human threshold.]

Without any pretense of having adequately surveyed the appropriate literature, two additional works in the fields of sociology and social psychology are cited to illustrate the current tendency to reach out for new contexts. Ernest Becker (1968) has written an ambitious essay on the unification of the science of man. His work is centered more in the history of philosophical and scientific thought than is the present volume or the other works cited. It offers, therefore, a different perspective on a similar set of issues. John Seeley (1967) is the peripatetic essayist of sociology just as Boulding is in the field of economics. His essays in a recent collection range over many of these issues from the perspective of social psychology.

Political Science

The field of political science is also becoming infected with the same germ. For illustrative purposes the work of a single author, Lindblom (1965, 1968), is cited.

In his writings Lindblom has focused on the political aspects of the process of social problem solving—aspects that have received short shrift in this volume. He is well known for his characterization of the political decision process in the Western world as a process of "disjointed incrementalism." (1965) He emphasizes that social problems are not truly solved but are subject to a never-ending series of attacks through the political problem-solving process. The process is incremental and deliberately exploratory. He sees the policy-making process as a "game of power," but one that incorporates a role for analysis and for cooperation as well as conflict resolution. There is much in the exposition to evoke the image of evolutionary experimentation developed in those pages. However, the process of "disjointed incrementalism" seems to be limited to the political nature of that aspect of evolutionary experimentation we characterize as normal problem solving. It does not successfully indicate how the political process can act to modify the rules of the game.

This bias is modified somewhat in his latest work, *The Policy-Making Process*. He may recognize this limitation when he says ". . . an established and stable play of power is an applecart that (a) would be upset by drastic and sweeping changes in the policies and that (b) has to be upset in order to achieve such changes" (p. 41). At the same time, Lindblom never adequately deals with the nature of what we characterize as paradigm shift, implying that, if it occurs, it emerges out of incremental changes of normal problem solving.

This bias and the heavy emphasis placed on the resolution of partisan conflict are a reflection of the work's basic historical and descriptive orientation. Further research, more directly guided by the metaphor of social learning, needs to consider ways in which the political process can move towards the political reorganization that is essential if the process of social learning is to be brought under social control.

The sketches and references offered in the preceding sections do not constitute a scholarly review, but they do offer impressive evidence of the degree to which many scientific disciplines are stimulated to reach out for new concepts. Many portals to the domain of social learning have been identified. One additional task remains before the mission of this review is complete.

Marxian Theory

In Marxist theory exists the most widely heralded theory of social change in the history of mankind. It has further provided the inspiration for the communist movement—surely one of the most comprehensive attempts at social action in social history. Since the process of social learning purports to be a theory of social change, it would be remiss if we did not offer a few words of comparison. To do so also offers the opportunity to emphasize in summary form several of the special characteristics of social learning.

In drawing these contrasts it is useful, first, to make a distinction not always clearly delineated in Marxist literature. Marxist theory is commonly described as historical or dialectical materialism. As expressed in the *Communist Manifesto* and *Das Kapital*, the essence of the theory is usually interpreted to mean the inevitable overthrow of capitalism—that the proletariat would necessarily revolt against its oppressors and create a classless communist society in which the state would wither away. This would take place through a rigid scheme of historical transformations that must pass through well-defined stages: primitive commune, slavery, feudalism, capitalism, socialism, leading to the communist classless society.

Modern Marxist scholarship is taking exception to the common interpretation, saying that the Marxist disciples—notably Lenin and Stalin—adopted a vulgarization of Marx's subtle philosophical thought that transformed it into a mechanistic determinism alien to Marx. This hardening of Marxist philosophy into dogma is said to have accompanied the bureaucracy, cult of personality, and intellectual isolation of the historical communist movement. (See Garaudy, 1967, and Avineri, 1968.) This view is based upon a reinterpretation of Marxist philosophy and social theory made possible by the discovery of the early writings of Marx (i.e., the *Economic and Philosophic Manuscripts of 1833 and 1844*).[7] These early writings present an image of Marx as the humanist rebelling against alienation who denied that Communism is "itself the goal of human development."

Marxist scholarship has not settled this matter. There are those that insist that there was a young Marx and an old—that the early philosophical humanist gave way to the Marx devoted to violent revolution and the "dictatorship of the proletariat."

It is not our role to settle the dispute of the specialists, but out of this background we can make several important observations. By the testimony of the modern Marxist scholars—some of them, like Garaudy, dedicated Marxists—the modern communist movement is guided by a mechanistic deterministic paradigm. It stands charged with committing the historical fallacy so much like the orthogenetic fallacy in the life sciences. It is committed to a terminal state teleology that it seeks to engineer from a transcendent base of sure knowledge

[7] These papers were unknown to Lenin or the Russian revolutionaries of 1917.

and technique. It is guilty of the pretension that it can short-circuit the process of social learning.

If the modern scholars are wrong and Marx did indeed fall into this trap in later life, he stands charged along with the communist movement. If we can accept the interpretation of the modern scholar or restrict our focus to the humanistic philosophy and social theory of the "young Marx," we can see that Marx had the genius to anticipate by more than a hundred years three of the most important aspects of a comprehensive theory of social learning.

First, Marx was concerned with the *process* of social change and had the insight to see it as experimental and evolutionary in form. This is revealed by Garaudy as he both interprets and quotes from Marx:

The central theme of historical materialism is this: "Men make their own history, but they do not make it just as they please; they do not make it under circumstances chosen by themselves, but under circumstances directly encountered, given and transmitted from the past." (From Marx, *The 18th Brumaire of Louis Bonaparte*, p. 15.)

Human history is qualitatively different from biological evolution but is rooted in it. . . . unlike animals, who adapt themselves simply to nature, man transforms it by his labor. . . . human environment is not a pre-existent, *given* nature but a nature always to some degree transformed, humanized. . . . But in satisfying his needs and creating technical means for satisfying them, man creates new needs. . . .

[Marxism is] the metamorphosis of speculative and dogmatic dialectics into a method of experimental research and discovery. . . . Nature obeys, lets itself be controlled. In acting on this hypothesis, I have power over nature. It is true that these hypotheses are self-destructive and that no single one of them could therefore claim to reveal the ultimate structure of being. But each dead hypothesis, because it has lived, has left us a legacy of a new power over nature. This power survives the old hypothesis, and the new one is the inheritor of the hypothesis it has replaced. These powers are cumulative and my activities today in utilizing these powers to control nature delineate at least a rough sketch of its structure, more and more [of which] is precisely known. (pp. 69, 96–97)

These statements can be interpreted as being consistent with the general concept of social learning.

Second, central to Marx's system of thought is his emphasis upon alienation. This is a word we have avoided in the text because it seems to drop a bit too readily from one's lips. It has almost become a part of the vernacular. In the process its coinage has been debased and, perhaps, is not quite able to carry the freight of understanding—particularly since it is applied to a variety of meanings. That, combined with the fact that emphasis upon the positive rather than the negative form seems preferable, inclines us to emphasize the idea of human development rather than the alienation that occurs when individual human potential is frustrated. Marx's concept of alienation grew out of his association with Feuerbach, one of the young Hegelians, who developed the concept of religious alienation. Marx transformed this concept into the concept of political and economic alienation.

For Marx, human alienation grows out of the fact that modern economic life transforms human beings into objects of an impersonal system where self-fulfillment is difficult if not impossible. This focus on human alienation growing out of the operation of social systems has the effect of recentering the aim of the social process upon the needs of man. This focus was lost in much of the writing of the nineteenth century and Marx was reacting strongly to it. As Becker (1968, p. 62) points out:

The idea of progress was thus "off-centered" in several ways and in several different countries—in England by Malthus, Spencer, and Darwin, and in Germany by Hegel. In France, Comte himself contributed to this off-centering by his own reliance on history, by objectifying it in his system. This gave "scientific historicism" wide currency, and the rest of the nineteenth century, as we know, went ahead full steam to try to "objectify" laws of historical progress. Ethnologists, sociologists, historians—all looked to reconstruct the stages of development of humanity in a single line. This would give them all an easy key to the secret of social dynamics.

In the process the needs of man become lost in the presupposition of a deterministic historical dynamic. It was certainly Marx's original intent to reject this historicism and in the process reestablish man at the center of the social process. The process is one that should be guided to serve man's needs.[8]

---

[8] In this connection, it appears that Popper (1934, 1945) may have levied the wrong criticism at Marx, even though an accurate one when applied to the com-

This also gives rise to the third attribute of Marx's thought that is similar to a comprehensive theory of social learning. It places emphasis on social action. Marx issued a call to activism on the grounds that a theory can demonstrate its worth only through practice. He wrote: "Man must prove [read, *test*] the truth, i.e., the reality and power ... of his thinking in practice. ... The philosophers have only *interpreted* the world differently, the point is, to *change* it."

In short, one might interpret Marx to have been saying that the social process should be one of evolutionary social experimentation aimed at fulfilling the fundamental needs of man through social action. These are certainly fundamental attributes of a comprehensive theory of social learning. The genius of Marx was to see and expound these important features of the social science paradigm that is currently emerging one hundred years later. This fact also explains the great appeal of Marxism and why it is still the metaphor that appeals to the young radicals of today. It is man-centered and activist. Marx transformed human redemption into a historical phenomenon capable of realization through human social action.

This raises several questions: Where did Marx go wrong? Why has the practice of these ideas created monolithic, deterministic, antihumanistic communist bureaucracies of today? Does this suggest that the emerging social science paradigm has, itself, a fundamental flaw?

The answer seems to lie in the fact that Marx was too far ahead of his time. He had the creative vision of a new metaphor and worked his whole life to give it concrete content. He had the misfortune to live at a time when neither science nor the accumulated experience of history could supply the kinds of knowledge essential to giving concrete and valid content to the concept. He also failed because of the pretension that such a theory and its concept are the work of one man-life. His social theory itself formed an evolutionary hypothesis, some elements of which have been tested through social action. In these concrete forms it has failed the test because the tests have demonstrated

---

munist movement and to the writings of Lenin and Stalin. Marx's whole system of thought grew out of a reaction to the prevailing historicism of the German philosophers. The limitations of Marxist theory become revealed in a different way, as we attempt to point out below.

goal divergence instead of goal convergence with respect to the humanistic goals. By comparing the more detailed content of the Marxist theory with the concepts of social learning articulated in this volume, we may be able to see more clearly the sources of the limitations of Marxist theory.

First, Marx was on the right track in his concern with social *process*, but he was able to develop only a limited concept of its nature. Marx's view of the social evolutionary process was still dominated by the idea of the Hegelian dialectic—the supposedly creative counter positioning of the thesis and antithesis, the idea of its opposite, action and opposing action. It pictured the process as wholly reactive in nature and dominated by conflict and its resolution. It made no place for the creative use of dialogue in social organization. Although Marx talks about "proving [testing] the truth" through social action in his early writings, he does not come to an explicit description of the nature of the process in terms of the reorganization of social systems. He does not deal explicitly with the nature of the social system as a behavioral entity so that the process can be formulated in these terms.

As a consequence, there is nothing in Marx that points up the distinction between evolutionary experimentation as normal problem solving and as a paradigm shift. Indeed, there is a serious imbalance in Marx because he sees social change exclusively as a process by means of which a paradigm shift is brought about. Marxist theory does not deal with the normal problem-solving process by means of which the controls of established social systems are made more consistent with their goals. With its emphasis upon alienation it neglects the deficit motives, or the requirements of material biological maintenance, and the importance of the day-to-day tasks of social maintenance including the social forms that organize them. It is assumed that this will somehow all come right if we can transform the goals of the system. The practicing communist systems have all had great difficulty in making these on-going production functions of society to "come right" and, where they have succeeded, they often have had to turn to other guides to social process and social organization than those supplied by Marxist theory.

The big thrust of Marxist theory is the reorganization of the total system. For Marx this has no generic meaning with respect to the boundary conditions of any system not already under superordinate control or subject to a change in the goals that such superordinate controls impose. It refers explicitly to the system of the nation state and it offers a single explicit developmental hypothesis about the means by which its behavior can be made more consistent with the goals of man.

This brings us naturally to the emphasis upon alienation. The second limitation of Marxist theory is the limited nature of its concept of alienation. Becker (1968, pp. 65–66) makes this point effectively: "Unfortunately, in his time Marx had to strive so hard to show that man's nature was social and historical that he could no longer approach man from within as Feuerbach had so well done. As a result, he could not proceed to formulate a truly complete theory of alienation, one that would rest upon a thoroughgoing phenomenology of individual striving."

Since human alienation was tied to the repressive character of early industrial organization, its amelioration was supposed to come from the abolition of private property. There was no behavioral science in Marx's day that could offer a better interpretation of the roots of human alienation. Certainly the association psychology of that time had nothing to offer. Marx cannot be blamed for not seeing that it is the growth motives or the developmental motives of the individual that must be served and that their satisfaction has little to do with social reorganization that abolishes private property. Indeed, the satisfaction of these motives, and the elimination of the alienation that arises when they are denied, must be largely served by the appropriate organization of the ongoing processes of society and the practice of normal problem solving—an aspect of the matter not recognized by Marx. We had to wait a hundred years for the beginnings of a developmental psychology to emerge.

Since the idea of process and the concept of alienation failed at the point where they had to be defined in operational terms, it is natural that the emphasis on social action had a biased character. Since alienation is associated with private property generating an exploited class, and the evolutionary process as a dialectic reactive process focused upon a total reformulation of the boundaries of the nation state (paradigm shift), social action takes the form of essential revolution to make a classless society through the abolition of property and establishment of the dictatorship of the proletariat. There is no

303

developed image of social action absorbed in evolutionary experimentation addressed to both normal and extraordinary problem solving leading to subsystem reorganization as well as occasional total boundary shifts. One cannot state that conflict forms of revolution may never be required to generate a boundary shift essential to bring about a convergence between social forms and human needs. If, however, such a revolution changes the locus of power without the new power élite having a realistic understanding of the processes by means of which the goals of the reform can be implemented, nothing is gained from the point of view of serving fundamental human goals. Only a transfer of power from one "in-group" to another takes place, as recent history has so ably demonstrated.

Marx was anxious for revolution to occur because he was afraid that man might lose forever the chance to control his own fate. His ideas were avidly seized upon because they offered the promise that man could realize his destiny through "good works" applied in the current historical period. The truth is that he offered an evolutionary hypothesis that was untestable through social action because it contained insufficient operational detail. The attempt to test it through the communist movement resulted in a vulgarization of the Marxist metaphor brought about by the necessities of practice (in the absence of such operational detail). Avineri (1968) suggests that Marx's theory may have been falsified by the very historical process he foresaw. The attempts to put Marx's ideas into practice have led to their denial.

Because it was not possible for Marx to specify adequately the operational characteristics of the social process, no test of it could be performed and the attempt to do so led instead to the testing of the old deterministic paradigm of social engineering. It seems possible in this era for social science to begin to specify more adequately the operational characteristics of social learning so that it can be progressively tested and revised through social action. We must not make the same mistake as Marx, however. This is the task for more than one man-life. It is the task of a mature, unified, and dedicated social science that operates to give orientation to social action as evolutionary experimentation—a social action that is conscious of its techniques and human goals and dedicated to their objective testing and reformulation through such action.

Appendix

A Note on the Concept of Resources

THE RESOURCE CONCEPT has always been one of major interest to the economist. The economist's preoccupation with deterministic optimizing systems performing the role of social maintenance has caused him most often to restrict the usage of the term to natural physical resources. It is the stock of these, and their associated flow of resource inputs to the social process, that is supposed to determine the optimum scale of the system.

As the economist has moved to consider the issues of economic development, we have seen that he has turned to an emphasis upon physical capital as a developmental resource. We commence to realize that the resource concept is more comprehensive and variable in form than we have been inclined to treat it. Our awareness of this fact is sharpened if we recognize that the content of the resource concept has undergone a succession of changes that are a product of the process of evolution. It is worth taking a brief space to review this transformation, to note its general characteristics, and to identify the special role of resources in development.

The most general definition of resources is provided by Webster's unabridged dictionary and suits our purpose admirably. Resources are defined simply and comprehensively as "available means." The term "means" implies an end or objective—a requirement or idea to be served. Resources are the available means for satisfying the requirement or embodying the idea. Since the ends of organism and cultures

305

become progressively modified in the course of evolution it is natural that the concrete content of the "available means" would change. Those attributes of nature that play the resource role become progressively modified. We seek to examine the gross dimensions of this progression.

One may even start with the evolutionary processes that antedate life processes. Modern theory holds that, just as the social process arose out of the operation of the biological process, life process arose out of the process of evolution of matter. It is held that the gaseous mass that once separated from the sun and formed the earth included heavy clouds of carbon that condensed and entered the primitive nucleus of the earth. There it combined chemically with the heavy metals at the core to form carbides during the cooling of the mass. These carbides came into contact with the superheated aqueous vapor of the atmosphere to produce the simplest organic compounds, hydrocarbons and all of their derivatives through oxidation. At the same time reactions with ammonia created a series of nitrogeneous derivatives. Thus, when the planet had cooled sufficiently to allow atmospheric condensation to form the first mantle of hot water, it contained organic substances in solution made up of carbon, hydrogen, oxygen, and nitrogen—precisely those compounds with the greatest chemical potentialities. As time went on these came to form colloidal systems that gradually assumed some differentiation of physicochemical structures that promoted colloidal stability and increased the efficiency of chemical absorption. Out of this colloidal process the "competitive spread of growth, struggle for existence and, finally, natural selection determined such a form of material organization which is characteristic of living things of the present time." (Oparin, 1938, p. 251.)

At the outset, then, the basic resources were the key inorganic elements yielding organic compounds—oxygen, carbon, hydrogen, and nitrogen. In time the organic compounds and inorganic compounds (like water) came to represent the basic resources in the formation of colloidal systems. These developed an assimilative capacity which was carried over into the primitive biological organisms that evolved a thermodynamic throughput based upon the assimilation of basic organic and inorganic compounds.

Under the influence of adaptive generalization the primitive organisms came to develop a more flexible assimilative capacity. In time this assimilative capacity came to include the assimilation of more primitive organisms that had already concentrated the organic and inorganic compounds through their behavior as organisms. In these more advanced organisms the basic resources became biological and reproducible. They still engaged in the direct assimilation of some nonbiological materials (chiefly water), but their access to material and energy resources became progressively indirect as the lower species came to serve as energy and material collectors in an assimilative chain.

At each stage in this evolution of organisms the "available means" applied *directly* to satisfying the organisms' requirements are redefined. The original resource components are not rendered unnecessary for the organism. Rather they become transmuted into more complex forms and more widespread options that serve the end organism by a more indirect path. Not only does this process have the effect of redefining the character of the resource applied directly to organism requirements, but it also enlarges the habitat or geographical range within which the organism can be supported. Not only does evolution redefine the "means," it also extends the "available." The more advanced organisms have more "degrees of freedom" as correlative mechanisms.

When we move from the biological phase into the cultural phase the stage is set for several more transformations of the resource concept.

In the first place, in the biological phase organisms are basically collectors. As the cultural phase advanced, organisms became basically husbanders. The organism applied its expanding knowledge of the environment to the conscious promotion, concentration, and control of reproducible biological resources (agriculture and animal culture). Progressively its principal resources came to be concentrated upon those animal and vegetable species that were most amenable to this form of control and concentration.

Following this the resource concept takes on a distinctly new turn. Once adaptive behavior evolves, an important part of the "available means" consists of the ideas embodied in conscious behavior that brings the earth's material and energy resources (inorganic, organic, and biological) to a controlled application to human needs. Thus, it happens that we have come to include "capital" in the resource concept. The essential concept of capital is the material embodiment of acquired ideas. Some primary material and energy resources are diverted from the immediate requirements to become embodied in artifacts and social forms that amplify the behavioral capacities of the human organism. All the artifacts of man that are not directly consumed constitute this new resource.

The forms of these artifacts have evolved along with mankind's understanding of the environment—i.e., his fund of relevant ideas. The embodiment of these ideas in material form comes to employ physical as well as biological materials. Thus, in the social phase, the direct assimilation of physical nonreproducible resources once again becomes important. Implicit in the development of the resource of artifact capital is the renaissance of material nonreproducible resources.

Because, individually, mankind is at the center of the cognitive grid he has tended to lose sight of the fact that human capital also constitutes an important resource (available means) in society. Basic to the means of satisfying requirements is the man who "knows how to do." He is the activating agent for the production and utilization of artifact capital and for the assimilation into society of all resources. In a sense he is a prior means.

In the biological phase the behavioral potential of an organism during its life cycle was largely fixed by the genotype—a given constant factor. Once acquired behavior could be accumulated in culture, the man who "knows how to do" becomes a product of culture. In a sense he becomes a cultural artifact—a basic means of accomplishing the ends of society (including his own). Thus, human resources, as applied to acquired ideas embodied in people, become an important part of the resource picture.

So we see that evolution has progressively redefined the material, energy and, ultimately, idea resources that serve as available means for biological and cultural organisms as it has transformed their organization and modes of behavior. The resource concept is seen as a relative concept. Whether an attribute of nature is actively or potentially assimilated by living organisms or social systems depends upon the requirements of its evolved behavioral mode. An attribute of nature that is relevant to behavioral system requirements, but beyond its effective behavioral range, is not a resource under this definition. Both changes in the ends to be served and changes in accessibility or availability bring about changes in the content of the resource concept.

We might offer this additional observation. Resources play a different role in the maintenance of life and in the transformation of life. A radiated terminated species or a steady state social system will have established a balanced, repetitive material-energy throughput where the role of the resources is stable and unchanging. It is our static state habit of mind and conventions of analysis that have tended at times to give to the resource concept a rather fixed content. But, as we have just seen, changing systems—physical, biological, or social—act to transform the content of the operative resource concept. And, indeed, it follows that if the ends that concern us are development and not maintenance, the resources that serve such ends must be interpreted to include a concept radically different from those that serve the end of self-maintenance.

In other words, we might project an additional step in the transformation of the resource concept. If social development serving human development becomes the goal, the ultimate resource, the basic means for achieving this end, is the ability to develop new ideas that improve the relationship between social system goals and controls operating in a mixed physical and social environment. We have just observed that as biological life and social culture have emerged, the constraints imposed at each stage by the restricted sources of energy, assimilative material, and behavioral know-how have been by-passed through the application of the new idea that extended the resource range. In a basic sense, then, the means for fulfilling the aims of development is the ability to generate new ideas in action. This creative capacity is the ultimate resource in meeting human objectives because it is evolving to the point of conscious and controlled application.

The resources that support social life, the means that serve our ends, are not just the physical material and energy inputs that maintain the essential thermodynamic throughput of the life processes. Even social maintenance requires human capital, or men who "know how to do." Social development requires, additionally, the activity of men and social organizations that "know how to learn" and who are able and willing to test what they learn against fundamental human values.

Bibliography

Abramovitz, Moses. "Resource and Output Trends in the U.S. since 1870." *Proceedings of the American Economic Association*, May 1956.

Adler, Mortimer J. *The Conditions of Philosophy*. New York: Atheneum, 1965.

Alexander, Samuel. *Space, Time, and Deity*. New York: Macmillan, 1920.

Allee, W. C.; Emerson, A. E.; Park, O.; Park, T.; and Schmidt, K. P. *The Principles of Animal Ecology*. Philadelphia: W. B. Saunders, 1949.

Allport, Gordon L. *Becoming: Basic Considerations for a Psychology of Personality*. New Haven: Yale University Press, 1955.

Argyris, Chris. *Understanding Organizational Behavior*. Homewood: Dorsey Press, 1960.

————. *Personality and Organization*. New York: Harper, 1957.

Arrow, Kenneth J. "The Economic Implications of Learning by Doing." *Review of Economic Studies*, vol. xxix, no. 80 (June 1962).

Ashby, W. Ross. "Principles of the Self-Organizing System." In *Principles of Self-Organization*, edited by H. Von Foerster and G. W. Zopf, Jr. New York: Pergamon Press, 1962.

————. *An Introduction to Cybernetics*. London: Chapman & Hall, Ltd., 1956. Reprint, New York: Wiley, 1964.

Avineri, Shlomo. *The Social and Political Thought of Karl Marx*. London: Cambridge University Press, 1968.

Bates, Marston. "Ecology and Evolution." In *Evolution After Darwin*, edited by Sol Tax, vol. i, Chicago: University of Chicago Press, 1960.

Becker, Ernest. *The Structure of Evil*. New York: George Braziller, 1968.

Benedict, Ruth. *Patterns of Culture*. Boston: Houghton Mifflin, 1934.

Bennis, Warren G., and Slater, Philip E. *The Temporary Society*. New York: Harper, 1968.

Bergson, Henri. *Creative Evolution*. New York: Henry Holt, 1911.

Berry, Brian J. L., and Pred, Allen. *Central Place Studies: A Bibliography of Theory and Applications*. Philadelphia: Regional Science Institute, 1961.

311

Bibliography

Bertalanffy, Ludwig von. *General System Theory: Foundations, Development Applications.* New York: George Braziller, 1968.
———. "The Mind-Body Problem: A New View." *Psychosomatic Medicine,* vol. xxiv (1964).
———. "General Systems Theory: A Critical Review." *General Systems Yearbook,* vol. vii, 1962a.
———. *Modern Theories of Development: An Introduction to Theoretical Biology.* New York: Harper, 1962b.
Boulding, Kenneth. *Beyond Economics.* Ann Arbor: University of Michigan Press, 1968.
———. "Economics and Ecology." In *Future Environments in North America,* edited by Fraser F. Darling and John P. Milton. New York: Natural History Press, 1966a.
———. *The Impact of the Social Sciences.* New Brunswick: Rutgers University Press, 1966b.
———. *The Meaning of the Twentieth Century.* New York: Harper & Row, 1964.
———. *The Image.* Ann Arbor: University of Michigan Press, 1961.
———. *The Organizational Revolution.* New York: Harper & Row, 1953.
———. *A Reconstruction of Economics.* New York: Wiley, 1950.
Bowman, Mary Jean, and Anderson, C. Arnold. *Education and Economic Development.* Chicago: Aldine, 1965.
Buhler, Charlotte. "Theoretical Observations about Life." *American Journal of Psychotherapy* 13 (1959): 508–581.
Burhoe, Ralph W. "Five Steps in the Evolution of Man's Knowledge of Good and Evil." *Zygon,* vol. 1, no. 2 (March 1967).
Burtt, E. A. *In Search of Philosophic Understanding.* New York: New American Library, 1965.
Bush, Robert R., and Estes, William K. *Studies in Mathematical Learning Theory.* Stanford: Stanford University Press, 1959.
Campbell, Donald T. "Evolutionary Epistemology." In *The Philosophy of Karl Popper,* edited by Paul A. Schlipp. Library of Living Philosophers. LaSalle, Ill.: Open Court Publishing Co., forthcoming.
———. "Variation and Selective Retention in Socio-Cultural Evolution." In *Social Change in Developing Areas: A Reinterpretation of Evolutionary Theory,* edited by H. R. Barringer, G. I. Blanksten, and R. W. Mack. Cambridge, Mass.: Schenkman Publishing Co., 1965.
———. "Common Fate, Similarity, and Other Indices of the Status of Aggregates of Persons or Social Entities." *Behavioral Science* 13 (1958): 14–25.
Cherry, Colin. *On Human Communication.* New York: Wiley, 1957.
Clark, Colin. *The Conditions of Economic Progress.* London: Macmillan, 1940.
Denison, Edward. *Why Growth Rates Differ.* Washington: Brookings Institution, 1967.

312

————. *The Sources of Growth in the United States.* Supp. report no. 3. New York: Committee on Economic Development, 1962.

Dewey, John. *Intelligence in the Modern World,* edited by J. Ratner. New York: Random House, 1939.

Dobzhansky, Theodosius. *Mankind Evolving.* New Haven: Yale University Press, 1962.

————. *Evolutionary Biology.* New York: Appleton-Century-Crofts, 1951.

Dunn, Edgar S., Jr., "A Flow Network Image of Urban Structure." *Urban Studies,* forthcoming.

————. *Review of Proposal for a National Data Center.* (Office of Statistical Standards, Statistical Evaluation Report no. 6). Washington: U.S. Bureau of the Budget, 1965.

Etzioni, Amitai. *The Active Society: A Theory of Societal and Political Processes.* New York: Free Press, 1968.

Fisher, R. A. *The Genetic Theory of Natural Selection.* Oxford: Clarendon Press, 1930.

Flavell, John H. *The Developmental Psychology of Jean Piaget.* Princeton: Van Nostrand, 1963.

Garaudy, Roger. *Karl Marx: The Evolution of His Thought.* New York: International Publishers, 1967.

Gerschenkron, Alexander. *Economic Backwardness in Historical Perspective.* Cambridge, Mass.: Harvard University Press, 1962.

Goodenough, Ward H. "Right and Wrong in Human Evolution." *Zygon,* vol. 1, no. 2 (March 1967).

Goudge, T. A. *The Ascent of Life.* Toronto: University of Toronto Press, 1961.

Greene, J. C. "Biology and Social Theory in the 19th Century: Auguste Comte and Herbert Spencer." In *Critical Problems in the History of Science,* edited by Marshall Clagett. Madison: University of Wisconsin Press, 1959.

Hagen, Everett E. *On the Theory of Social Change.* Homewood: Dorsey Press, 1962.

Harris, Britton. "The City of the Future: The Problem of Optimal Design." *Regional Science Association Papers,* vol. 19 (1967).

Harris, Errol E. *The Foundations of Metaphysics in Science.* New York: Humanities Press, 1965.

————. *Revelation through Reason.* New Haven: Yale University Press, 1958.

Heidegger, Martin. *Existence and Being.* Chicago: Regency Press, 1949.

Henderson, Lawrence. *The Fitness of the Environment.* New York: Macmillan, 1927.

Huxley, Julian. *Evolution: The Modern Synthesis.* New York: Wiley, 1964.

————. "Evolution: Cultural and Biological." *Yearbook of Anthropology,* 1955.

Ittelson, William H., and Cantril, Hadley. *Perception: A Transactional Approach.* New York: Random House, 1954.

Klausner, Samuel Z. "Man and His Non-Human Environment." Mimeographed. Washington: Resources for the Future, 1968.

———. *The Study of Total Societies*. New York: Doubleday, 1967.

Kluckhohn, Clyde. "The Scientific Study of Values and Contemporary Civilization." *Proceedings of the American Philosophical Society*, vol. 102, no. 5 (October 1958).

Kroeber, A. L. "Evolution, History, and Culture." In *Evolution After Darwin*, vol. 2. Chicago: University of Chicago Press, 1960.

Kuhn, Thomas S. *The Structure of Scientific Revolutions*. Chicago: University of Chicago Press, 1964.

Langer, Susanne K. *Mind: An Essay on Human Feeling*, vol. i. Baltimore: Johns Hopkins Press, 1967.

Lasuén, José-Ramón. "Growth Poles and Regional Planning: The Venezuelan Case." Unpublished manuscript. Washington: Resources for the Future, 1969.

Likert, Rensis. *The Human Organization: Its Management and Practice*. New York: McGraw-Hill, 1967.

Lindblom, Charles E. *The Policy-Making Process*. New York: Prentice-Hall, 1968.

———. *The Intelligence of Democracy*. New York: Free Press, 1965.

Lorenz, Konrad. "Gestalt Perception as Fundamental to Scientific Knowledge." *General Systems Yearbook*, 1962.

Lotka, Alfred J. *Elements of Physical Biology*. Baltimore: Williams & Wilkins, 1924. Reprint, New York: Dover Publications, 1956.

Lowry, Ira S. "Comments." *Regional Science Association Papers*, vol. 19 (1967).

McClelland, David C. *The Achieving Society*. Princeton: Van Nostrand, 1961.

Mack, Ruth P. *Planning on Uncertainty*. New York: Wiley, forthcoming.

Maslow, Abraham. *Toward a Psychology of Being*. New York: Van Nostrand, 1968.

Matson, Floyd W. *The Broken Image*. New York: George Braziller, 1964. Reprint, Doubleday Anchor, 1966.

Matson, Floyd W., and Montague, Ashley. *The Human Dialogue: Perspective on Communication*. New York: Free Press, 1967.

Mayr, Ernst. *Animal Species and Evolution*. Cambridge, Mass.: Harvard University Press, 1963.

———. "The Emergence of Evolutionary Novelties." In *Evolution After Darwin*, edited by Sol Tax, vol. i. Chicago: University of Chicago Press, 1960.

Mead, George Herbert. *Mind, Self, and Society*. Chicago: University of Chicago Press, 1934.

———. *Philosophy of the Act*. Chicago: University of Chicago Press, 1932.

Meier, Gerald M. *Leading Issues in Development Economics*. New York: Oxford, 1966.

Morgan, C. Lloyd. *Emergent Evolution*. New York: Henry Holt, 1923.

Morison, Robert S. "Science and Social Attitudes." *Science*, vol. 165, no. 3889 (11 July 1969).

Murphy, Roy E., Jr., *Adaptive Processes in Economic Systems*. New York: Academic Press, 1965.

Narall, Raoul S., and Bertalanffy, Ludwig von. "The Principle of Allometry in Biology and the Social Sciences." *General Systems Yearbook*, 1956.

Nelson, R. R.; Peck, M. J.; and Kalachek, E. D. *Technology, Economic Growth, and Public Policy*. Washington: Brookings Institution, 1967.

Nirenberg, Marshall. Editorial. *Science*, 11 August 1967.

Oparin, A. I. *The Origin of Life*. New York: Macmillan, 1938.

Peirce, Charles S. "Search for a Method" (1893). Reprinted in *Collected Papers of Charles Sanders Peirce*. Cambridge, Mass.: Harvard University Press, 1958.

Pelz, Donald C., and Andrews, Frank M. *Scientists in Organization*. New York: Wiley, 1966.

Perloff, Harvey S.; Dunn, Edgar S., Jr.; Lampard, Eric E.; and Muth, Richard F. *Regions, Resources, and Economic Growth*. Baltimore: Johns Hopkins Press, 1960.

Polanyi, Michael. "Life's Irreducible Structure." *Science*, vol. 160 (21 June 1968): 1308–1312.

Popper, Karl R. *The Poverty of Historicism*. 3d ed. New York: Harper, 1964.

―――. *The Open Society and Its Enemies*. London: Routledge & Kegan Paul Ltd., 1945. Reprint, New York: Harper, 1963.

―――. *The Logic of Scientific Discovery*. New York: Basic Books, 1959. First published in Germany as *Logik der Forschung*, 1934.

Rensch, Bernhard. *Evolution Above the Species Level*. New York: Wiley, 1959.

Rostow, W. W. *The Stages of Economic Growth*. Cambridge: Cambridge University Press, 1960.

Sackman, Harold. *Computers, System Science, and Evolving Society*. New York: Wiley, 1967.

Sahlins, M. D., and Service, E. R. *Evolution and Culture*. Ann Arbor: University of Michigan Press, 1960.

Schon, Donald A. *Technology and Change*. New York: Dell, 1967.

―――. *Displacement of Concepts*. London: Tavistock, 1963. Republished as *Invention and the Evolution of Ideas*. London: Social Science Paperbacks, 1967.

Schultz, Theodore William. *The Economic Value of Education*. New York: Columbia University Press, 1963.

Seeley, John. *The Americanization of the Unconscious*. New York: International Science Press, 1967.

Sellars, Roy Wood. *The Philosophy of Physical Realism*. New York: Macmillan, 1939.

Simon, Herbert. "The Architecture of Complexity." *Proceedings of the American Philosophical Society*, vol. 106, no. 6 (December, 1962).

Simpson, George Gaylord. "Biology and the Nature of Science." *Science*, vol. 139, no. 3550 (January 1963).

———. *The Major Features of Evolution*. New York: Columbia University Press, 1953.

———. *The Meaning of Evolution*. New Haven: Yale University Press, 1949.

Slobodkin, L. B. "A Game Theoretic Interpretation of Biological Evolution." In *Population Biology and Evolution*, edited by R. C. Leworstim. Syracuse: Syracuse University Press, 1968.

Smuts, Jan Christian. *Idealism and Evolution*. New York: Macmillan, 1926.

Solo, Robert A. *Economic Organizations and Social Systems*. Indianapolis: Bobbs-Merrill, 1967.

Stanley-Jones, Douglas, and Stanley-Jones, K. *The Kybernetics of Natural Systems*. New York: Pergamon Press, 1960.

Stephens, Benjamin H. "Location Theory and Programming Models: The Von Thunen Case." *Regional Science Association Papers*, vol. 20, 1968.

Swanson, Guy E. Book Review. *Science*, vol. 162 (20 December 1968): 1376–1377.

Teilhard de Chardin, P. *Man's Place in Nature*. New York: Harper & Row, 1966.

———. *The Phenomenon of Man*. London: Balius, 1959.

Tillich, Paul. *Systematic Theology*, 3 vols. Chicago: University of Chicago Press, 1951, 1957, 1963.

Toynbee, Arnold. *Study of History*. London: Oxford University Press, 1946.

Waddington, C. H. *The Ethical Animal*. London: Allen & Unwin, 1960.

Werner, Christian. "The Role of Typology and Geometry in Optimal Network Design." *Regional Science Association Papers*, vol. 20, 1968.

Whorf, Benjamin Lee. *Language, Thought and Reality*. Cambridge, Mass.: Massachusetts Institute of Technology Press, 1956.

Wiener, Norbert. *Cybernetics: On Control and Communication in the Animal and the Machine*. New York: Wiley, 1948.

Index

Abramovitz, Moses, 14n
Absolute values, 171–74, 176–77, 185
Activity network: Social system as, 204–7, 210; transformation of, 217–20, 247
Adaptability: Destruction of, 231; need for, 123; in organisms, 45, 50, 75–76, 78n, 164–65; progressive trend in, 9, 64–65, 73; in social realm, 85, 94–96, 101–4, 178, 228. *See also* Behaviorai change
Adaptation, anticipatory, 99, 110, 118, 131, 208, 214
Adaptive generalization, 42, 45–52, 60, 62, 64, 73, 306–7; effect of, 76; in individuals, 275; as paradigm shift, 147–50 (*see also* Paradigm shifts); preconditions for, 54–56, 97, 129, 148–49, 175; progressive trend in, 64–68; relationship to specializing adaptations, 57; in sociogenesis, 89–96, 107n, 148, 158
Adaptive radiation. *See* Adaptive specialization
Adaptive specialization, 42–48, 50–51, 54–57, 60, 73, 178; at ecosystem level, 62, 64, 68n; as evolutionary trap, 45, 100–101; in independent populations, 86–89; in individuals, 275; as normal problem-solving, 147 (*see also* Normal problem-solving); in social subsystems, 91–94
Adler, Mortimer J., 260, 271
Advanced societies, problem of social organization in, 214–17

Agriculture, as social subsystem, 92, 102n
Alexander, Samuel, 271
Alienation, Marxist theory of, 300, 302–3
Allee, W. C., 68n
Allometric growth: Biologic, 53; models, 14–17, 22, 26, 119, 145
Allport, Gordon L., 108n, 178–79, 181, 260, 274
Analysis: Need for new methodologies in, 256n; in physical science, 240–41; structural, 47–49, 131, 145n, 189. *See also* Concept models; Economic Analysis
Analyzing behavior, 105, 108–11, 207–9
Anderson, C. Arnold, 289
Andrews, Frank M., 225, 231, 280
Anomalies, in social organization, 121, 140, 143, 146, 149, 221, 246, 260, 263–64, 277
Anthropology, 33–34, 86–88
Argyris, Chris, 280
Arrow, Kenneth J., 285
Artifacts: Increased complexity of, 79, 90, 95, 308; individuals as, 182
Ashby, W. Ross, 21n, 27, 276–77
Atomic power, 104, 215
Authority, in social system, 171, 173–74, 290
Avineri, Shlomo, 298, 304

Backward economies. *See* Underdeveloped (backward) economies
Bates, Marston, 60–61, 63, 68n
Bayesian theorem, 279

317

Index

Entropy transfer, 277

Environment: In biological evolution, 41, 44–54, 59–61, 69–70, 73; control of, 99–100, 105, 116–19, 155, 264, 299; "fitness" of, 65–67, 292; human cognition of, 108; "operational," 60–62, 110–11; "potential," 60–63, 69; in social evolution, 82–90, 95, 98, 102, 239; threats to, 104–5

Epistemology, 268–72

Equilibrium: In biological evolution, 45, 62, 200; in economic analysis, 17–20, 32–33, 121–22, 162, 200, 223, 276, 279; in psychological theory, 178–79, 274; in social systems, 206, 251

Estes, William K., 71n

Etzioni, Amitai, 281, 294

Europe, culture in, 87

Evaluation of behavior, 108–9, 160, 162, 175, 238, 240, 242; inversion of, 172; need for, 134, 252, 260; scientific resistance to, 161–63, 170, 264; in system development, 282

Evolution: Chemical, 97, 100, 306; genetic, 40–41, 50–56, 59, 77, 80; narrative explanations in, 127–28, 248n; neo-Lamarckian theory of, 76n; of nervous system, 48–51, 53, 76; novelties in, 45, 52, 54–55, 59, 73; progressive, 42n, 64–68, 89–90, 95–96, 98; resources in, 306–8; of social values, 104, 163–67, 169, 242; thresholds in, 97–100, 148–49, 262–64; traps in, 45, 100–105, 154; trends in, 126–27. See also Biological evolution; Social change (evolution)

Evolutionary epistemology, 268–71

Evolutionary experimentation, 111–60, 209, 212, 238, 240–42, 260; developmental hypotheses in, 163, 169, 228, 242, 244, 247–48, 250; efficiency of, 154–59, 169–73; goals in, 163, 166–67, 173–85, 190 (see also Goals, in social development); hazards of, 172–73; informational needs, 256; in Marxist theory, 302; and personality development, 275; purposive use of, 244–51, 304. See also Social learning

Evolutionary hypotheses, 246, 301, 304. See also Developmental hypotheses

Evolutionary phenomena, and scientific laws, 125–33, 155, 250

Existentialism, 177

Exogenous changes, in problem-solving, 142n

Experience, sharing of, 77–78, 80, 98

Experimentation, social, 158. See also Evolutionary experimentation

Export base, growth of, 17

External economies, 142

Extinction, in phylogenesis, 45, 57, 84, 101, 103

Fechner, G. T., 274

"Feelings," role of, 270

Feuerbach, L. A., 300, 303

Fisher, R. A., 53

Flavell, John H., 275

Flexibility. See Adaptability

Forecasting, 71, 115–25, 130–32, 145n. See also Prediction

"Forgetting," 272

Freud, Sigmund, 274

Friedrich, W. G., 122n

Game theory, 70–71, 132–33, 143, 146, 162, 279

Garaudy, Roger, 298–99

Garrison, W. L., 122n

General equilibrium model, 17, 19–20, 122n, 162, 200, 279

General systems theory, 275–78, 294

Genetic: Drift, 128–29; engineering, 215; evolution, 40–41, 50–56, 59, 77, 80

Gerschenkron, Alexander, 131, 286

"Gestalt" perception, 270

Goals, in social development: In decision theory, 280; defined, 161n; effect of differences in, 158–59; evaluation of, 163; evolution of, 104, 249; in Marxist theory, 302–4; need for concern with, 190, 244–46, 251, 255–56; need for integration with control systems, 216–17, 221–22, 228–29, 243–44, 246; in normal problem-solving, 211–12, 242; in paradigm shifts, 140–41, 213, 216–17; reification of, 172–73, 203, 216; role in creative process, 83–85, 108–11, 133–36, 159–85, 208–9, 241–42, 254; scientific resistance to dealing